IET HISTORY OF TECHNOLOGY SERIES 20

Series Editor: Dr B. Bowers

The Early History of Radio

from Faraday to Marconi

Other volumes in this series:

The Early History of Radio

from Faraday to Marconi

G.R.M. Garratt MA CEng FIEE FRAeS

The Institution of Engineering and Technology in association with The Science Museum, London

Published by The Institution of Engineering and Technology, London, United Kingdom

First edition © 1994 The Institution of Electrical Engineers
Reprint with new cover © 2006 The Institution of Engineering and Technology

First published 1994
Reprinted 2006

The Institution of Engineering and Technology
Michael Faraday House
Six Hills Way, Stevenage
Herts, SG1 2AY, United Kingdom

www.theiet.org

British Library Cataloguing in Publication Data
A catalogue record for this product is available from the British Library

ISBN (10 digit) 0 85296 845 0
ISBN (13 digit) 978-0-85296-845-1

Printed in the UK by Short Run Press Ltd, Exeter
Reprinted in the UK by Lightning Source UK Ltd, Milton Keynes

Contents

Figures and Diagrams

Figures 1, 2, 3, 4, 6, 7, 9, 10 are reproduced by courtesy of the Trustees of the Science Museum, and figures 11, 12, 13, 14, 15, 16 by courtesy of the Marconi Company Ltd.

Gerald Garratt (1906-1989)

Foreword

The idea of communicating instantly with people at a distance has always appealed, and the electric telegraph made it possible in the mid nineteenth century. Until then messages could travel no faster than a horse on land or a sailing ship at sea. Admittedly there were a few semaphore telegraphs which could send a message from one station to another, provided it was not too far away and in the hours of daylight. The telegraph made it possible to send messages anywhere and at any time, but it depended on an insulated wire, expensive to install and easy to damage. Even while the telegraph wires were spreading round the world, a new technology was evolving. Faraday and Maxwell conceived the idea of electromagnetic waves; Hertz showed that they could be created and detected; Marconi began the task of harnessing them.

Much has been written about Faraday and about Marconi. Radio from the time of Marconi onwards has also been the subject of many studies. Gerald Garratt had a special interest in the sequence of development from Faraday's first concept of the electromagnetic field, through the theoretical and experimental investigations of numerous workers, to Marconi's practical application. In this book he traces the 'pre-history' of radio - the steps between Faraday and Marconi. He was especially interested in the work of Hertz, and one of the highlights of his Museum career was getting to know Frau Elizabeth Hertz and her daughters after they left Nazi Germany for Cambridge in 1936; this enabled him to acquire some of Hertz' original manuscripts for the Science Museum.

Gerald Garratt was born in 1906 and educated at Marlborough College and Caius College Cambridge, where he missed a year following a motorcycle accident. Despite that mishap he learnt to fly on an Avro 504, and became a founder member of the Cambridge University Air Squadron. In 1928 he was commissioned into the RAF Reserve of Officers as a Pilot Officer and remained in the Reserve until he retired as Wing Commander in 1966.

On leaving Cambridge he was apprenticed at Metropolitan Vickers, then worked at the Royal Aircraft Establishment, Farnborough, on gyroscopic instruments. He took part in some of the early experiments in 'blind landing'. During the Second World War he was Senior Engineer Staff Officer with the Royal Indian Air Force.

Most of his career, apart from wartime service, was in the Science Museum which he joined as an Assistant Keeper in September 1934. For many years he was in charge of the Communications Collections. He wrote a number of papers on the history of telecommunications. His book, *One hundred years of submarine cables,* was published in 1950. He was also an enthusiastic amateur radio operator, G5CS, with his own transmitter and beam aerial at home, and he was responsible for setting up the Museum's radio station, GB2SM, in 1955.

Outside the Museum he had varied interests. He built himself three Lotus cars from kits, kept eighty hives of bees, was the local agent for bee-keeping equipment and honorary secretary of the local hospital's League of Friends. He was a lover of classical music, especially that of Mozart.

The author completed the first six of the seven chapters in this book before his death. He intended to round off the story with a chapter introducing Marconi. The author's daughter, Susan, has edited her father's text and prepared a final chapter on Marconi based on a paper published by the Institution of Electrical Engineers and a lecture her father gave at the Institution in 1972, at a meeting organized jointly with the Radio Society of Great Britain. The publishers are grateful to her for preparing this material for publication, believing, with her, that this account of the very early history of radio will be of interest to many.

Brian Bowers
Series Editor
October 1993

Chapter 1

Introduction

Whether it be due to a genuine quest for knowledge or merely to a desire to satisfy passing curiosity, there can be no doubt as to the frequency with which one hears the question 'Who was the inventor of so-and-so?' Unfortunately, such questions seldom admit of concise answers, the truth being that very few inventions are the work of single individuals. On the contrary, even quite simple inventions are generally the outcome of a chain of prior developments which have been spread, not infrequently, over a considerable period of time and to which a number of individuals have separately contributed.

Never was this more true than in the field which forms the subject of this book, for although there is a widespread belief that the invention of wireless telegraphy was the work of a young Italian, Guglielmo Marconi, and although it is perfectly true that the first patent ever to be granted for a system of wireless telegraphy stands in his name, the fact is that his achievement was only the practical application of scientific developments and discoveries which had been spread over a period of almost eighty years.

Marconi's success - and let us not for one moment belittle his achievement - was largely due to the fact that his work was based upon the very sure foundations which had been laid in earlier years by some of the leading scientists and mathematicians of the nineteenth century. Outstanding among them are the names of Faraday, Maxwell, Lodge and Hertz and it is the purpose of this book to review briefly the parts which these and others played in preparing the ground for Marconi and for the practical development of what we now call 'Radio Communication'.

In essence, the early history of wireless telegraphy - or perhaps it would be more correct to call it the 'pre-history' - is the history of electromagnetic waves, the prediction of their existence by Faraday, the mathematical definition of the conditions for their propagation by Maxwell, and the eventual demonstration of their physical existence by Hertz with his experimental confirmation of the identity of their characteristics with those of visible light.

We are apt to speak familiarly of electromagnetic waves perhaps without pausing to reflect that an electric wave can have no existence unless there is a corresponding magnetic wave, and that no electric current can flow without the creation of a magnetic field. Yet the knowledge of this inter-dependence of electric and magnetic phenomena, so commonplace today that we accept it almost unconsciously, was not won easily. Indeed, during the early years of the last century, the nature of the relationship was completely unknown. Voltaic electricity itself was a new field and, although it was widely suspected that there was some connection between the forces of electricity and those of magnetism, the real basis of the relationship remained a mystery. As the true facts began to emerge during the 1820s, the realization that electric and magnetic forces did not act along straight lines was gradually to bring about a revolution in the understanding of physical forces and Faraday's share in the early development of what we now call 'Field Theory' must count as one of his major contributions to scientific progress.

The genesis of electromagnetic theory

In order to set the discoveries and theories of the nineteenth century in their true perspective, it is desirable to review very briefly the 'state of the art' during the early years of the century. The various phenomena of 'static' or 'frictional' electricity were widely familiar, the Leyden jar had been invented in 1745 and the frictional machine had become established as a fairly reliable generator of electricity at about the same time. There were even several proposals for the use of static electricity as the basis for systems of telegraphy, the last of these being that of Francis Ronalds in 1816. But low-pressure, 'voltaic' electricity was quite a new subject - Volta's Pile of zinc and silver discs, the first electric battery, was only invented in the year 1800 and although various forms of chemical cell quickly followed, with their ability to provide a continuous source of electricity, the range of phenomena which thereby became demonstrable was so different from the familiar phenomena of static electricity that virtually a new branch of science had appeared. The concept of electrical potential and current had as yet to be appreciated - volts and amps were as yet unknown - and, of course, no means existed for the measurement of electrical quantities.

By the year 1820 it had already been realized that some connection almost certainly existed between the forces of electricity and magnetism - both, for example, showed properties of attraction and repulsion - but the real connection remained obscure until Hans Christian Oersted discovered that the passage of an electric current through a wire could deflect a magnetic needle. Oersted had devoted his thoughts for a number of years to the probability of some connection between electricity and magnetism but his actual discovery was made, almost by accident, during the course of a lecture he was giving in the University of Copenhagen. The effect he observed was a feeble one and, with remarkable restraint, he refrained from announcing his discovery until he had had time to confirm it by more detailed experiments.

At last he was ready and in July 1820 he published the memoir, written in Latin, which was to broadcast the news of his discovery and render his name immortal in the history of science.[1] Apart from the basic facts of his discovery, there were two sentences in the concluding paragraphs of his memoir which were to prove of special significance:

> It is sufficiently evident from the preceding facts that the electrical conflict is not confined to the conductor but is dispersed quite widely in the surrounding space. It is also evident that the forces of this electrical conflict operate in circles, for in the absence of such an assumption, it would seem impossible that the current in the wire when placed below the magnetic needle could turn it towards the east and when placed above towards the west.

The announcement of Oersted's discovery caused a sensation throughout the scientific world of the time, not so much on account of the discovery of the magnetic effect caused by the current, since this had been expected in some form or other for a number of years, but on account of the quite extraordinary nature of the magnetic force it produced. The fundamental basis of Newtonian science had been built upon the assumption that forces act only along straight lines between two points but here were forces operating in circles; a completely new, unexpected and inexplicable phenomenon!

Another immediate and important consequence of Oersted's discovery was the invention by Schweigger of his 'multiplier', or 'galvanometer', as it soon came to be called. Observing that the deflection of a magnetic needle by a current passing through a wire above it was the same as that produced by the return current passing through a wire beneath, he arranged the wire by bending it so that the current passed first above and then below the needle. As he expected, he obtained a larger deflection and, by giving the wire yet further turns, the effect was still further multiplied. Thus, before the end of 1820, the first measuring instrument was available.

The news of Oersted's discovery soon reached Paris where it was communicated by François Arago to the members of the French Academy on the 11 September 1820. Among the audience on that occasion was the Professor of Mathematics at the Ecole Polytechnique, André Marie Ampère, who was particularly astonished at the revelation since he had been for years firmly among the ranks of those who believed that no interaction between the forces of electricity and magnetism was possible. Ampère at once proceeded to repeat Oersted's experiment and within a few days he had greatly elaborated it by showing that there was a comparable reaction between two adjacent current-carrying wires. Indeed, within a short period he was able to establish all the elementary laws of electrodynamics (a word which he coined) and, by his subsequent mathematical analysis, to establish the whole subject on a firm theoretical basis.

Ampère, however, was in a quandary. He could not deny the facts of Oersted's discovery which his own experiments had so amply confirmed but he seems to have been unwilling to abandon completely his belief that reaction between the forces of electricity and magnetism was impossible. In his anxiety to reconcile his former conviction that electrical and magnetic forces were of a

Figure 1   André-Marie Ampère

totally different nature, he developed a theory that the properties of a magnet were due to the continual circulation of intermolecular currents in the magnet's body. An electric current flowing in an external wire reacted, therefore, not with the magnetism of the magnet but with the intermolecular currents which caused the magnetism. An electric current, according to Ampère, could react with another electric current but not with a magnet, and he was thus compelled to invent the concept of intermolecular currents to explain Oersted's discovery of the reaction between a current-carrying wire and a magnetic needle!

The weakness of such a theory may seem very evident today but it cannot have been so obvious to Ampère's contemporaries in the early spring of 1821. Apart from the fact, however, that the existence of continual intermolecular currents would have implied that a magnet must always be a little warmer than its surroundings, the theory inevitably demanded a fundamental re-thinking of the whole theory of matter. While Ampère's theory might satisfy the mathematicians, it demanded far more from the chemists and it was precisely this aspect of the matter which was to trouble Faraday when he came to repeat Oersted's experiments later in the summer.

References

1   H. C. Oersted, *Experimenta circa effectum Conflictus Electrici in Acum Magneticam* (Experiments in the Effect of a Current of Electricity on the Magnetic Needle), Copenhagen, 1820. English translations were published in *Annals of Philosophy*, **16**, 1820, p.276, and in the *Journal of the Society of Telegraph Engineers*, 1877, **5**, pp.459-73. For a photographic reproduction of the original, see Bern Dibner *Oersted and the Discovery of Electro-Magnetism,* Burndy Library, Norwalk, Conn., 1961.

Chapter 2

Faraday

Michael Faraday (1791-1867)

Tributes have often been paid to Michael Faraday as the outstanding pioneer of the electrical industry, a reputation which is perhaps largely founded upon his discovery of electromagnetic induction in 1831. It has been far less common, however, to regard Faraday as the scientist upon whose work the whole theory and practice of radio communication has developed. Yet such is the case, for without Faraday there would have been no Clerk Maxwell, no Hertz, no Marconi, and it is almost certain that the advent of wireless telegraphy would have been delayed by many years. Faraday's work was in fact fundamental to the development of electromagnetic theory and to the propagation of electromagnetic waves and it is therefore appropriate here to review briefly the stages by which, over a period of more than thirty years, his ideas and theoretical concepts developed to form the foundations for the work of Clerk Maxwell.

Faraday was not a trained scientist; he was the son of a blacksmith, almost entirely self-taught and he had no mathematical ability whatever. With such an unpromising background, it is not altogether surprising that he was regarded by many of his contemporaries as an unreliable heretic whose views were to be treated with the greatest reserve. Later generations were perhaps too ready to look upon Faraday as an intuitive empiricist or merely as an able experimenter - but perhaps one can scarcely blame them when even a mathematical physicist of the stature of Clerk Maxwell was regarded with scepticism by the great majority of his contemporaries.

In reality, Faraday was a daring theorist who was ever prepared to disagree with convention and established beliefs if the circumstances and experiment seemed so to demand. With the benefit of perspective which the passage of more than one hundred years has conveyed, we can now see Faraday as one of the most brilliant scientists of the nineteenth century. He was certainly a master of experimental technique but he also made greater contributions in the realm of theoretical physics than any of his contemporaries. It was his work which laid the foundations for the development of classical field theory and in doing so he prepared the way for the development of practical radio half a century later.

When Faraday first learned of Oersted's discovery from Sir Humphry Davy, he had already been Davy's assistant at the Royal Institution for more than seven years. His interest had been mainly in the chemical and electrochemical fields and at the time when the news reached him in October 1820 he was heavily preoccupied with an investigation into the properties of iron alloys and the manufacture of steel. Faraday's father had been a blacksmith and the practical knowledge of iron working which he must have gained in his early years would have been a very useful background to the scientific investigation which, as an analytical chemist, he had been invited to undertake.

Besides the time-consuming steel investigation, Faraday had two other contemporary interests, one being long-term research into the properties and compounds of chlorine and the other his courtship of Sarah Barnard whom he was to marry in June 1821. As Professor Williams has remarked, 'it seems fair to conclude that while he found Oersted's discovery an exciting one, he had not the time to peruse it for himself'.[1]

Whatever the reason, it seems clear that, although Faraday and Davy repeated and confirmed Oersted's experiment almost as soon as the news reached them from Paris in October 1820, they both failed to appreciate at once the extraordinary conflict with which the concept of a circular magnetic force opposed traditional beliefs. Oersted had shown quite clearly that the magnetic force was a circular one but this aspect of the experiment seems to have been entirely overlooked by both Faraday and Davy and for many months they both remained under the impression that the force between the wire and the needle was one of simple attraction. Curious though this misunderstanding may be, it is clear from the almost complete absence of references to electromagnetism in Faraday's diary for the period that he had not attributed any great importance to Oersted's discovery and, his mind occupied with more immediate matters, he had not given the subject serious thought.

Faraday's serious work on electromagnetism commenced during the summer of 1821 as a result of an invitation to contribute an historical article on the subject to the *Annals of Philosophy*.[2] Faced with this task, he prepared for it by repeating the experiments performed by Oersted, Ampère and others and, in doing so, he at once realized the error into which he had fallen the previous autumn. He now realized that the magnetic force associated with a current-carrying wire was of a circular nature and did not consist of simple attractions and repulsions as he had earlier assumed.

Had Faraday been an academically trained mathematical physicist he might have found himself seriously disturbed by this realization but, uninhibited as he was by traditional beliefs and Newtonian philosophy, he saw nothing very remarkable in a force which appeared to act along a curved path, and there can be little doubt that his later conception of 'lines of force' had its origin in the experiments he carried out during the summer of 1821.

FARADAY
1791 - 1867

Figure 2 Michael Faraday

Although it is not strictly relevant to the subject of this book, there was one other development from Faraday's work at this period which was to have such a profound influence on the whole future development and application of electricity that a brief mention seems desirable. In repeating the experiments of Oersted and Ampère, Faraday had not been content with a simple repetition to demonstrate an effect but had carried out a lengthy series of experiments and devoted very considerable thought to the facts observed. Having thereby formed a full appreciation of the circular nature of the magnetic force around the wire, he argued that there should be a tendency for a wire to describe a circular path around a magnetic pole. To demonstrate this effect, Faraday set up a vertical magnet with its upper pole projecting above the surface of a trough of mercury. A wire was arranged above with its lower end free to revolve in the mercury around the magnetic pole. When a current was passed through the wire, it revolved around the pole, thereby demonstrating the conversion of electricity into mechanical work and constituting the first electric motor.[3]

During the ten years which followed his discovery of electromagnetic rotation, Faraday was far too busy with the duties of his office and with other researches to devote any sustained effort to the problems of electricity and magnetism. There was his metallurgical work to complete, his researches into the properties and compounds of chlorine and, from 1827, he was committed to an intensive research into the manufacture of optical glass. Yet in spite of these preoccupations, it is evident that the subject of electromagnetism was never far from his mind although the few references which appear in his diary for the period render it none too easy to follow the evolution of his thoughts which, in 1831, were to lead him to his epoch-making discovery of electromagnetic induction.

It may well be, as Professor Williams has suggested, that Faraday's thoughts were kept simmering by his continuing scepticism and questioning of Ampère's theory as to the causes of magnetism. There can be little doubt that his ideas were stimulated by the discovery in 1825 by François Arago and Samuel Christie that the rotation of a copper disc could produce a corresponding motion in a magnetic needle freely suspended above it. It is more than likely that his thoughts were further stimulated by the experiments of Professor Moll of Utrecht on powerful electromagnets, but it is still, at least to the present writer, something of a mystery as to exactly what it was that suddenly inspired the brief but intensive period of experimentation in the summer of 1831, the results of which were to have such a profound influence over the whole future development of electrical science.

It had been known for ten years that an electric current gives rise to an effect corresponding to magnetism and there were many besides Faraday who had argued that the converse should also be true - that magnetism should give rise to electricity. As Faraday remarked:

> it appeared very extraordinary, that as every electric current was accompanied
> by a corresponding intensity of magnetic action at right angles to the current,

good conductors of electricity, when placed within the sphere of this action, should not have any current induced in them, or some sensible effect produced equivalent in force to such a current.

These considerations, with their consequence, the hope of obtaining electricity from ordinary magnetism, have stimulated me at various times, to investigate experimentally the inductive effect of electric currents.

The above extract is taken from the introductory paragraph of the paper[4] which Faraday read to the Royal Society on 24 November 1831 in which he gave a detailed account of the long series of experiments he had recently carried out and in which, at last, the mystery was cleared up. As Faraday now showed, the earlier experiments had failed because the essential factor of *change* had either been missing or had escaped observation. Quoting from the same paper:

Several feet of copper wire were stretched in wide zig-zag forms, representing the letter W, on one surface of a broad board; a second wire was stretched in precisely similar forms on a second board, so that when brought near the first, the wires should everywhere touch, except that a sheet of thick paper was interposed. One of these wires was connected to a voltaic battery, the other with the galvanometer. The first wire was then moved towards the second, and *as it approached*, the needle was deflected. Being then removed, the needle was deflected in the opposite direction . . .

At last the explanation was clear; it required a *change* in the magnetic state to induce an electric current in a nearby conductor, and in the course of the experiments which he performed following this discovery, Faraday established all the fundamental laws of electromagnetic induction.

Professor Williams has suggested that Faraday's epoch-making discovery may have been inspired by the work of Fresnel who had, a few years previously, shown that the phenomenon of light was due to a wave motion and, by so doing, he had demonstrated that means existed for the transfer of energy without the transfer of tangible matter. To the present writer, however, it seems improbable that Faraday's discovery of electromagnetic induction could have been directly inspired by Fresnel's theory. While it is undoubtedly true that Faraday had been greatly impressed by Fresnel's work, the connection seems too remote. It is more likely that the discovery of electromagnetic induction arose directly from Faraday's acute observation in the course of his many experiments with electricity and magnetism.

The series of experiments which Faraday carried out during 1831 and which led to his discovery of electromagnetic induction was to have a sequel of some significance in relation to the subject of this book. While it would be unwise to attach too much historical importance to the incident, it sheds an interesting light upon the personal relationships in the world in which Faraday moved and provides a remarkable insight into the extraordinary intuition with which Faraday was endowed.

During the course of his experiments, he had formed the belief that magnetic force did not act instantaneously upon a distant magnet but that an appreciable element of time was required for the transmission of the force. By analogy, he found it possible to suppose that charges due to static electricity also required time for their transmission and, developing the idea, he could conceive that such forces were propagated by a kind of wave motion comparable 'to the vibrations upon the surface of disturbed water'.

It seems quite impossible that Faraday could, in fact, have detected the very short time of transmission we now know to have been involved. With the limited apparatus and techniques which were available to him, such precision of measurement would have been altogether beyond his reach and although one might be tempted to credit his ideas to chance and to guesswork, the revolutionary nature of his proposition, his own stature as an experimental physicist and the extraordinary accuracy of his prediction all seem to justify the thought that his belief was yet another mark of the intuition and genius which was to enshrine him among the leading physicists of the nineteenth century.

Faraday lived in an age when men of science were sometimes peculiarly sensitive in matters relating to priority of discovery or invention and, some years previously, he had unwittingly become involved in a situation where he was suspected, quite unjustly, of having tried to appropriate to himself the credit belonging to another. The incident was a trivial one of no possible consequence but it left a deep mark in the mind of one of such scrupulous integrity and he determined henceforth to run no risks and to safeguard his name against any similar accusation. He may seem to us to have been over-sensitive but one must remember that he was a man of comparatively humble origin, he was largely self-educated and there were many among his scientific contemporaries who regarded him with feelings of resentment and jealousy. In the circumstances, he was probably wise not to make public a tentative theory for which he was unable to provide an experimental proof. To have done so at this stage in his career would undoubtedly have involved him in criticism and ridicule but he was, nevertheless, reluctant to surrender the chance to claim credit for his theory or to run the risk of others seeing his experiments and depriving him thereby of his claims to priority.

He therefore decided to record his views in a sealed document which, on 12 March 1832, he handed to one of the Secretaries of the Royal Society by whom, a month later, it was deposited in the Society's Strong Box where it remained, unopened, for more than one hundred years!

It was at last decided that sufficient time had elapsed to examine the contents of the packet and, accordingly, the seal was broken on 24 June 1937 by Sir William Bragg, the then President of the Royal Society, in the presence of the Council. In this somewhat romantic manner Faraday's ideas, expressed in his own handwriting, came to light for the first time:

Royal Institution
March 12th 1832

Certain of the results of the investigations which are embodied in the two papers entitled 'Experimental Researches in Electricity' lately read to the Royal Society, and the views arising therefrom, in connexion with other views and experiments lead me to believe that magnetic action is progressive, and requires time, i.e. that when a magnet acts upon a distant magnet or piece of iron, the influencing cause (which I may for the moment call magnetism) proceeds gradually from the magnetic bodies, and requires time for its transmission, which will probably be found to be very sensible.

I think also, that I see reason for supposing that electric induction (of tension) is also performed in a similar progressive way.

I am inclined to compare the diffusion of magnetic forces from a magnetic pole to the vibrations upon the surface of disturbed water, or those of air in the phenomenon of sound; i.e. I am inclined to think the vibratory theory will apply to these phenomena as it does to sound, and most probably to light.

By analogy, I think it may possibly apply to the phenomenon of induction of electricity of tension also.

These views I wish to work out experimentally; but as much of my time is engaged in the duties of my office, and as the experiments will therefore be prolonged, and may in their course be subject to the observation of others, I wish, by depositing this paper in the care of the Royal Society, to take possession as it were of a certain date; and so have right, if they are confirmed by experiments, to claim credit for the views at that date; at which time as far as I know, no one is conscious of or can claim them but myself.

M. Faraday

It would be altogether wrong to attach too much historical significance to this fascinating document or, for example, to regard Faraday as the prime inventor of radio on the grounds that it refers to wave propagation. Nevertheless, in the light of Faraday's later contributions to what has become known as field theory, one may at least remark on the fact that knowledge of the propagation of electromagnetic waves can now be traced to these 'Original Views' which Faraday deposited with the Royal Society in 1832.

Joseph Henry (1797-1878)

In order to preserve some semblance of chronological order, it is necessary to break off at this point to refer briefly to the work of Joseph Henry who had been appointed Professor of Natural Philosophy at Princeton College, New Jersey, in 1832 and whose discovery that the sudden discharge of a Leyden jar was of an oscillatory nature was an event of great significance.

In the course of a long series of experiments on electromagnetic induction, in which he employed the discharge of Leyden jars in place of the galvanic battery as used by Faraday, Henry showed that exactly analogous effects were obtained, currents being induced in a secondary circuit which were capable of magnetizing a steel needle, if a small helix surrounding the needle were included in the circuit.

It had been known for a number of years that if a Leyden jar was discharged through a wire which passed at right angles near to a steel needle, the needle became magnetized, but it had often been observed that the direction of magnetism thus induced in the needle appeared to be capricious and to depend on the distance of the needle from the wire or on the magnitude of the Leyden jar discharge. This extraordinary anomaly seemed to offend against common sense and, in Henry's words, 'appeared to be at variance with all our theoretical ideas of the connection between electricity and magnetism'.

Henry, whose methods and achievements as an experimentalist closely paralleled those of Faraday, performed a vast number of experiments to determine the cause of this anomalous magnetization. He found that when he magnetized a *thick* needle by discharging a Leyden jar through a spiral of wire surrounding the needle, the needle was always magnetized in the same direction, but that if he used *thin* needles, they were sometimes magnetized in one direction and sometimes in the other. It was well-known that a given needle had only a certain capacity for receiving magnetism, a thick one more than a thin one, and that, beyond a certain value, no increase in the magnetizing current could produce any permanent increase in the magnetism of the needle. Henry argued that the phenomenon of anomalous magnetization could only be explained by assuming that the discharge of the Leyden jar was oscillatory, the initial flow of electricity in one direction being followed by a lesser flow in the reverse direction and so on, until all the energy in the discharge was dissipated.

In the case of the thick needle, Henry argued, even the strong current in the initial swing of the discharge was insufficient to saturate the magnetic capacity of the needle and although the magnetism produced by the initial flow of the current would be reduced by the reverse flow in the succeeding swing, it would not be cancelled altogether. The needle, therefore, always retained a remnant of the magnetism induced by the initial wave of current and was, therefore, always magnetized in the same direction. Using thin needles, however, when magnetic saturation was being reached, it might well happen that the flow of current in the reverse swing was sufficient to cancel completely the magnetism produced by the initial discharge. In such a case the thin needle would retain a permanent magnetism in the opposite direction to that of the thick one.

Reporting on the results of his experiments in 1842 to the American Philosophical Society, Henry wrote:

This anomaly, which has remained so long unexplained, and which, at first sight, appears at variance with all our theoretical ideas of the connection of electricity and magnetism, was, after considerable study, satisfactorily referred by the author to an action of the discharge of the Leyden jar which has never before been recognized. The discharge, whatever may be its nature, is not correctly represented by the single transfer of an imponderable fluid from one side of the jar to the other; the phenomena require us to admit *the existence of a principle discharge in one direction, and then several reflex actions backward and forward, each more feeble than the preceding, until the equilibrium is obtained.* All the facts are shown to be in accordance with this hypothesis, and a ready explanation is afforded by it of a number of phenomena which are to be found in the older works on electricity, but which have until this time remained unexplained.[5]

Arising out of the same series of experiments, Henry made a further observation of considerable significance in regard to the distance at which the inductive effects associated with the discharge of a Leyden jar can be detected. Quoting from the same paper of 1842, Henry wrote:

a remarkable result was obtained in regard to the distance at which inductive effects are produced by a very small quantity of electricity; a single spark from the prime conductor of the machine, of about an inch long, thrown on the end of a circuit of wire in an upper room, produced an induction sufficiently powerful to magnetize a needle in a parallel circuit of wire placed in the cellar beneath, at a perpendicular distance of thirty feet with two floors and ceilings, each fourteen inches thick, intervening.

By these experiments, therefore, Henry had established two important facts:

(a) the oscillatory nature of the discharge of a Leyden jar and
(b) the considerable distance at which inductive coupling between two circuits could be effective.

While neither discovery had any immediate application, they both formed a part of the background against which the techniques of electromagnetic wave propagation were beginning to unfold.

Faraday's lines of force

A comprehensive survey of Faraday's concept of lines of force and of the stages by which, over a period of more than thirty years, his early ideas developed to form the basis of electromagnetic field theory would occupy a considerable volume. In this brief review it is possible only to sketch very superficially the steps by which the concept gradually took form to serve as the very foundation upon which Clerk Maxwell was to build his classical theory.

As was suggested earlier it seems probable that the seeds of Faraday's lines of force were sown in 1821 when he repeated the experiments of Oersted and Ampère and realized that the magnetic force in the vicinity of a wire carrying a galvanic current followed a *circular* path around the wire. Had Faraday's earlier education followed more formal lines it is likely that he would have found it difficult to accept the evidence of his own experiments and sought, as did Ampère, to find some other explanation for an observed fact which seemed so outrageously at variance with traditional beliefs. Fortunately, Faraday's mind was not fettered with preconceptions and he saw no need at all to adopt a complicated and somewhat improbable theory merely for the purpose of explaining the experimental facts in terms of a traditional theory. As far as Faraday was concerned the forces were circular - and that was that!

Familiar as we are today with all the simple manifestations of electricity and magnetism, it is perhaps not easy to realize that one hundred and fifty years ago, these simple relations were all matters of the deepest mystery. Oersted had discovered a link between magnetism and electricity - but what was magnetism? What, indeed, was electricity? Did it partake of the nature of a fluid? Or two fluids? What was the connection, if any, between what they then called 'common electricity' (which we would now call 'static' or 'high-voltage' electricity) and the 'galvanic' electricity derived from Volta's pile or from chemical cells?

Faced with these and similar problems, it is not surprising that an experimental physicist of Faraday's calibre should have found himself fascinated by the whole subject or that, in particular, he should have devoted special attention to the nature of the forces between magnets and wires carrying electric currents. Very gradually throughout the 1820s it became increasingly clear to him that, whatever the phenomenon of magnetism might be due to, and in whatever peculiar state the particles of a magnet might be, the forces associated with that state extended outwards from the magnet and into the space in its vicinity. The nature of the forces which he had but vaguely pictured in 1821 were gradually becoming clearer, but it was not until his discovery of electromagnetic induction in 1831 that he first referred explicitly to the concept of lines of force in a footnote:

> By magnetic curves, I mean the lines of magnetic forces which would be depicted by iron filings; or those to which a very small magnetic needle would form a tangent.[6]

The 'electrotonic' state

During the course of the experiments which led up to his discovery of electromagnetic induction, Faraday formed a tentative belief that a conductor in the vicinity of a magnet or of a current-carrying wire was in some special or peculiar condition to which he gave the name 'electrotonic state'. He confessed, however:

> The peculiar condition shows no known electrical effects while it continues; nor have I been able to discover any peculiar properties exerted, or properties possessed, by matter while retained in this state.[7]

This extract has been taken from the paper which he read to the Royal Society on 24 November 1831 but it is not at all clear from the context just what it was which led Faraday to postulate a state of matter, the peculiar condition of which was totally undetectable! He seems to have been allowing himself the indulgence of giving expression to a vague and half-developed theory and, as one could see later, he was making the mistake of supposing that it was the *conductor* which, under the influence of a magnet, was in a peculiar condition instead of realizing that it was, in fact, the *space* in the vicinity of the magnet which was in the peculiar condition.

Within only a few weeks, however, his ideas began to clarify and in his second paper[8] to the Royal Society which he read on 12 January 1832 he virtually abandoned the concept of an 'electrotonic' state:

> Thus the reasons which induced me to suppose a particular state in the wire have disappeared: and although it seems to me unlikely that a wire at rest in the neighbourhood of another carrying a powerful electric current is entirely indifferent to it, yet I am not aware of any distinct *facts* which authorize the conclusion that it is in a particular state.

Very gradually, one concept was giving way in his mind to another: the 'electrotonic state' was giving way to what we would now call the magnetic field.

Lines of force in motion

Yet another concept of great significance had its origin at this time - the idea that a magnetic field was not necessarily to be regarded as a stationary affair but that in certain cases:

> the magnetic curves themselves must be considered as moving . . . from the moment at which they begin to be developed until the magnetic force of the current is at its utmost; expanding as it were from the wire outwards . . . On breaking the battery contact, the magnetic curves (which are mere expressions for arranged magnetic forces) may be conceived as contracting upon and returning towards the failing electric current . . .

Here for the first time we have a clear suggestion that magnetic forces may be considered as being propagated *in time* and extending progressively throughout the neighbouring space. In view of the historic interest of this concept of propagation, the reader may care to note that the paper[9] from which the above extract has been taken is roughly contemporary with the sealed and secret document already referred to on page 11; it was actually written about ten weeks later.

Electrostatic induction

Prior to these events in the early months of 1832, it is broadly true to say that philosophers had tended to concentrate on the 'end effects' on bodies in consequence of the forces between them; in future the manner in which such forces were transmitted, whether through empty space or through the medium of appreciable matter, was to become the predominant problem. During the course of his experiments on magnetism and electromagnetic induction throughout the 1820s and early 1830s, Faraday had become more and more firmly convinced that the forces associated with such phenomena required some form of 'medium' for their transmission. He found it increasingly difficult to believe that such forces could possibly be transmitted through completely empty space - in the presumed manner of gravitational forces - particularly since all his experiments seemed to indicate that magnetic and electromagnetic forces often acted along a curved path. It seemed impossible to imagine that any force could act along a curved path unless it operated through some form of medium with properties comparable, at least, with conventional matter.

But Faraday was in a quandary. He realized that his belief that some form of medium was essential to the conveyance of magnetic forces was utterly at variance with the accepted theory of 'action at a distance', but to challenge a theory which had the support of the great majority of contemporary scientists was little short of heresy. Faraday saw the need to 'tread warily', to produce incontrovertible evidence, step by step, until his case was overwhelming - and the next stage was clearly to prove that the induction of static electrical charges was also dependent upon the existence of a medium.

In the paper[10] which forms the Eleventh Series of his *Experimental Researches in Electricity* and which he read to the Royal Society on 21 December 1837, Faraday proved conclusively that 'electric induction is an action of the contiguous particles of the insulating medium or di-electric'. His experiments showed clearly that the induction of electrical charges ordinarily involved forces which acted along curved lines and, in an unusually direct challenge to the traditionalists, he wrote:

> I do not see how the old theory of action at a distance and in straight lines can stand, or how the conclusion that ordinary induction is an action of contiguous particles can be resisted.

Although the words of this extract carry a sense of conviction which runs through the remainder of the paper, it is evident that, for a good many years, Faraday still retained lingering doubts as to the exact manner in which forces acted through space. The old controversy was far from settled and, even fourteen years later, we find him writing:

> How the magnetic force is transferred through bodies or through space we know not; whether the action is merely action at a distance, as in the case of gravity; or by some intermediate agency . . . Such an action may be a function of the ether; for it is not unlikely that, if there be an ether, it should have other uses than simply the conveyance of radiation.[11]

In fact Faraday was still convinced that some form of medium was involved and it was perhaps only the absence of any firm proof which left any trace of doubt in his mind.

The ether 'topsy turvy'

It may be of interest to divert at this stage to refer briefly to the theories and controversies which raged from the sixteenth century onwards around the phenomena of gravitation and light and the means whereby such phenomena were transmitted through space. Not least among the problems which faced the early philosophers were those which sought an explanation of the mysterious connection between the motion of the moon and the rise and fall of the tides or of the means whereby the light and heat of the sun could traverse empty space to comfort the earth.

Prior to the time of Newton it is broadly true to say that the theories of René Descartes (1596-1650) were widely accepted. Briefly, his teaching implied that bodies can *only* act upon each other by direct pressure or impact and the 'Cartesians', as his followers were called, thus denied the possibility of 'action at a distance'. In order to explain, for example, the transmission of light through empty space, they assumed that space was a *plenum,* filled with a medium - the ether - which, though imperceptible to human senses, was nevertheless capable of transmitting forces and exerting effects upon material bodies immersed in it. They taught that the ether consisted of particles, continuously in motion in chains of vortices and that it was the motion of these vortices which enabled the ether to perform its mysterious functions.[12]

Cartesian conceptions were overturned - but only very gradually - by the work of Isaac Newton (1642-1727), whose theory of universal gravitation was published in his *Principia* in 1687. Initially, however, the reception given to Newton's ideas was decidedly lukewarm and there were very few who accepted wholeheartedly his theory that all bodies attract each other with a force proportional to the product of their masses and inversely proportional to the square of the distance between them. To many, his laws seemed to violate that accepted philosophic principle of the time that 'matter cannot act where it is not' and, curiously, even Newton himself did not seem to realize that his theory amounted to a denial of the Cartesian 'plenum of vortices' and to an acceptance of action at a distance. It was a long while before a belief in universal gravitation became firmly established - continental philosophers were particularly slow to adopt it - and during the first quarter of the eighteenth century an acceptance of Newton's theory was, in Lord Kelvin's words, 'a peculiar insularity of our countrymen'.

By the end of the eighteenth century, however, the revolution was complete and the idea of action at a distance had become so firmly established that the mere idea that gravitational forces - or, indeed, electric or magnetic forces - required any form of medium for their propagation seemed almost as unthinkable as action at a distance had seemed to Newton and his contemporaries one hundred years earlier.

To Faraday, however, almost alone among his contemporaries, it seemed improbable, to say the least, that forces could act through a complete vacuum where no matter or medium of any sort existed. As a practical man, Faraday was always disposed to look for practical explanations and he found himself unable to accept the idea of action at a distance, even though its use by the mathematicians always seemed to lead to the right solutions. He might not understand *how* forces could be transmitted through a medium, or indeed, of what such a medium might consist, but of the basic fact that magnetic and electric forces partook of the nature of strains and tensions in some form of medium he was firmly convinced. Clerk Maxwell was to share his views, and, in time, was to bring about the final rejection of action at a distance - but before dealing with the work of Maxwell, we must briefly refer to Faraday's further development of his concept of lines of force.

Back to Faraday's lines of force

Not long after the completion of his researches into static electricity in which, as we have already seen, he had shown that the phenomenon of electrostatic induction was an action dependent upon the nature of the dielectric, Faraday's health began to deteriorate and for a number of years - 1839-1844 - he was virtually unable to concentrate or to undertake serious work. It was as though he had 'burned himself out' by twenty years of intensive physical and mental exertion and a long rest was needed even partially to restore his powers.

With his gradual return to health, Faraday began to resume his duties at the Royal Institution and his former custom of delivering lectures from time to time at the regular Friday evening meetings. At one of these in 1846, when he was called upon to speak at short notice, he ventured to speculate on 'Ray-Vibrations' and the manner in which they might be transmitted through space and matter.[13] Confessing to his audience that he was merely throwing out 'the vague impressions of my mind', he enquired:

> whether it was not possible that the vibrations which, in a certain theory, are assumed to account for radiation and radiant phenomena may not occur in the lines of force which connect particles, and consequently masses of matter together; a notion which, so far as it is admitted, will dispense with the ether which, in another view, is supposed to be the medium in which these vibrations take place.[14]

Faraday was here suggesting - very tentatively - that radiation might be due to vibrations along the lines of gravitational forces. He did not really believe in the existence of an ether but this was only one of the problems to which mid-nineteenth-century science was unable to provide an answer and in the absence of a satisfactory alternative, a measure of speculation was inevitable.

From the time of his earliest experiments in electricity and magnetism during the autumn of 1820, Faraday had been accustomed 'to think and speak of lines of magnetic force as representations of the magnetic power; not merely in the points of quality and direction, but also in quantity'. He had made use of the concept throughout virtually the whole of his scientific career but in 1851, as he approached sixty years of age, he felt that the time had arrived when the basic concept should be thoroughly examined and the real nature of lines of force accurately determined. In particular he began to ponder on the question as to whether the lines of force had a real physical existence, and in what one can only describe as a spate of papers in 1852, he examined their characteristics in the greatest detail and concluded by declaring his belief in the reality of their existence.

There were peculiar differences between electrostatic and magnetic lines of force; the ends of an electrostatic line always terminated on different bodies, one positively charged, the other negatively, and the strain, if strain it was, appeared to traverse a curved path of polarized particles. The ends of a magnetic line of force, on the contrary, always terminated on the same body - but did they 'terminate' at all? Were they not, in fact, continuous throughout the body of the magnet forming uninterrupted loops?

These and many other similar problems Faraday investigated experimentally during 1851 and he described his findings in the long paper[15] which he read to the Royal Society towards the close of the year. In this paper, the twenty-eighth in the series of *Experimental Researches in Electricity*, Faraday examined the whole concept of lines of force, their characteristics and their distribution both within a magnet and throughout the neighbouring space. He followed up the Royal Society paper with two more popular addresses to the Royal Institution. In the first of these[16] he described his methods of detecting and plotting the lines of force and the experiments which proved that magnetic lines of force were closed curves which permeated the body of a magnet. In the second,[17] he dealt with the different characteristics of the forces exerted by gravity, light, electricity and magnetism, and while he expressed his belief in the physical existence of lines of force, he concluded by suggesting that 'it is, I think, an ascertained fact that ponderable matter is not essential to the existence of physical lines of force'.

Faraday's final paper[18] on magnetism and lines of force was a lengthy survey in which he dealt with almost every known facet of the subject. Declaring that the paper contained much of a speculative and hypothetical nature, he had to confess that the solution was not yet in sight:

> With regard to the great point under consideration, it is simply, whether the lines of magnetic force have a *physical existence* or not? . . . I think the existence of *curved* lines of magnetic force must be conceded; and if that be granted, then I think that the physical nature of the lines must be granted also.

There is no doubt that Faraday believed in their real existence. He regarded them as strains, but strains in what? He had rejected the ether; he simply did not believe that an all-pervading homogeneous fluid could possibly possess the properties required, but what remained? He had come very close to giving the answer 'space' but although he did not cease for a number of years to give further thought to all the problems posed by the concept of lines of force, the time had now come to hand on the torch to a young mathematician, James Clerk Maxwell.

References

1. L. Pearce Williams, *Michael Faraday: a biography*, Chapman & Hall, London, 1965.
2. M. Faraday, 'Historical Sketch of Electromagnetism', *Annals of Philosophy*, 1821, **2** (new series), pp.195-200, 274-290. Continued in **3**, pp.107-121.
3. M. Faraday, 'On some new Electro-Magnetical Motions and on the Theory of Magnetism', *Quarterly Journal of Science*, 1821-2, **12**, pp.74-96; 'Description of an Electro-Magnetical Apparatus for the Exhibition of Rotary Motion', *Quarterly Journal of Science,* 1821-2, **12**, pp.283-285.
4. M. Faraday, 'On the Induction of Electric Currents and on the Evolution of Electricity from Magnetism', *Philosophical Transactions* (hereafter *Phil. Trans.*), 1832, **122**, pp.125-162; *Experimental Researches in Electricity* (hereafter *ERE*), **1**, pp.1-139.
5. Joseph Henry, 'On Induction from Ordinary Electricity; and on the Oscillatory Discharge', *Proceedings of the American Philosophical Society*, 1841-3. **2**, pp.193-l96; reprinted in *Scientific Writings of Joseph Henry,* Smithsonian Miscellaneous Collections, 1887, **30**, pp.200-203.
6. See ref.5, p.154.
7. *ibid.*, p.161.
8. M. Faraday, 'On Terrestrial and Magneto-Electric Induction', *Phil.Trans.,1832*, 122, pp.163-194; *ERE* **2**, pp.140-264.
9. *ibid.*
10. M. Faraday, 'On Induction', *Phil.Trans.*, 1838, **28**, pp.1-40, 79-81; *ERE* **11**, pp.1161-1317.
11. M. Faraday, 'On Lines of Magnetic Force', *Phil. Trans.*, 1852, **142**, pp.25-56; *ERE* **28**, pp.3070-3176.
12. Sir Edmund Whittaker, *A History of the Theories of Aether and Electricity,* **1**, Thomas Nelson 1951.
13. The occasion was 3 April 1846. Many writers have assumed wrongly that the speaker who defaulted was Charles Wheatstone, because Faraday also described some of Wheatstone's latest work. In fact, the planned speaker was James Napier. See Brian Bowers, *Sir Charles Wheatstone*, 1975, p.22, and David Gooding and Frank A. J. L. James, *Faraday Rediscovered*, 1985, p.149.
14. M. Faraday, 'Thoughts on Ray Vibrations', *Philosophical Magazine* (hereafter *Phil.Mag.*), 1846, **28** (93rd sries), pp.345-350.
15. M. Faraday, 'On Lines of Magnetic Force, their definite character and their distribution within a Magnet and throughout Space', *Phil.Trans,* 1852. **142**, pp.25-56; *ERE* **28**, pp.3070-3176.
16. M. Faraday, 'On the Lines of Magnetic Force', *Proceedings of the Royal Institution*, 1851-4, **1**, pp.105-108.
17. M. Faraday, 'On the Physical Lines of Magnetic Force', *Proceedings of the Royal Institution*, 1851-4, **1**, pp.216-220.
18. M Faraday. 'On the Physical Character of the Lines of Magnetic Force'. *Phil. Mag.* 1852, **3** (4th series), pp.401-428.

Chapter 3

Maxwell

James Clerk Maxwell (1831-1879)

Maxwell's interest in electromagnetic theory was almost certainly inspired initially by another young Cambridge wrangler, William Thomson (later Lord Kelvin), who in 1846-47 had made a mathematical investigation into the similarities between electromagnetic phenomena and elasticity. In the course of this investigation, he had examined the equations of equilibrium of an incompressible elastic solid in a state of strain and he had shown that elastic displacement was analogous to the distribution of electric forces in an electrostatic field. Thomson's memoir[1] was concerned with equilibrium conditions but its results were such as to suggest to Maxwell a few years later that the analogies between elastic strain and electrostatic forces might be extended and applied to the propagation of electromagnetic forces through an appropriate medium. One of the first steps in following this line of thought was to investigate the possible characteristics of an appropriate medium and, having been greatly impressed with his reading of Faraday's *Experimental Researches in Electricity*, and with his whole concept of lines of force, Maxwell commenced by translating Faraday's ideas into mathematical terms.

Maxwell was only twenty-four when he read his first great paper, 'On Faraday's Lines of Force',[2] to the Cambridge Philosophical Society in 1855. The paper, which filled fifty-six pages, not only subjected Faraday's ideas to most rigorous mathematical analysis but it linked them firmly with the analogies which had been suggested by Thomson. There was nothing fundamentally new in this first paper but by expressing and analysing Faraday's theories in a language intelligible to mathematicians, it made them accessible to and acceptable by a much wider range of readers.

It is perhaps important to emphasize that in the 1850s there were still two schools of thought. The mathematicians and physicists constituted the vast majority, and they regarded the forces between charges of electricity or between magnetic poles as due to direct action at a distance. To the members of this school, the space between such charges was exactly the same as any other space, the only relevant factor being the distance. Upon this hypothesis they solved many problems correctly, the solutions often being conspicuous for the elegance and ingenuity of the analytical methods employed. In the other school, with lamentably few supporters, was Faraday who, almost alone for nearly thirty years, had been advancing the theory that the electric charges or magnetic poles were but the starting points for the series of lines of force which spread out through the surrounding space. In his view, it was these lines, of whatever they might consist, which were responsible for the effects produced by the electric charges and by the magnets.

The importance of Maxwell's first paper, 'On Faraday's Lines of Force', lay in that, by translating Faraday's theories into mathematical form, he was able to show that they led to exactly the same numerical solutions as the theory of action at a distance. Previously the mathematicians had been contemptuous, regarding Faraday as an untutored heretic. 'I declare', wrote G. B. Airy, the Astronomer Royal, in 1855, 'that I can hardly imagine anyone who knows the agreement between experiment and calculations based upon the theory of action at a distance to hesitate an instant in the choice between this simple and precise action on the one hand and anything so vague and varying as lines of force on the other'. But now, with Maxwell's demonstration that the mathematicians and Faraday both arrived at the same answer, the situation was greatly changed.

During the five or six years which followed the publication of Maxwell's paper, while its contents were being absorbed, little of consequence transpired but Maxwell himself must have been giving a great deal of thought to the nature of a medium which could serve as a vehicle for the transmission of electric and magnetic forces. By this time, it had been generally accepted that light consisted of transverse vibrations in an intangible but all-pervading medium and that this medium must possess a small but real density. Inasmuch as the medium appeared capable of transmitting motion with a great, but not infinite, velocity, it appeared to possess properties analogous to mass and elasticity, and these necessarily implied the existence of kinetic and potential energies.

By about 1860, Maxwell had become impressed with the similarities between the phenomena of light and those of electromagnetism and he had become convinced by Thomson's arguments that Faraday's experiment showing the rotation of the plane of polarization of a beam of light in a magnetic field afforded strong evidence that the phenomenon of magnetism possessed the characteristics or tendencies of rotation. While, therefore, he suspected that light itself was of an electromagnetic nature, it was not possible to confirm this without first investigating in detail the characteristics and properties of the medium - the problematical ether - by means of which it was supposed to be transmitted. With this aim in mind, Maxwell developed his ideas on the possible structure of a medium which would account for all the observed phenomena - for magnetic attraction, the action of electric currents upon each other, and electromagnetic induction. The outcome of all this work was published in a series of papers[3] in the *Philosophical Magazine* during 1861-2.

Figure 3   James Clerk Maxwell

Observing that a magnetic field appeared to consist of a tension along the lines of force together with a pressure at right angles to the line, Maxwell posed the question: 'What mechanical explanation can we give of this inequality of pressures in a fluid or mobile medium?' He went on to suggest that:

> The explanation which most readily occurs to the mind is that the excess of pressure in the equatorial direction arises from the centrifugal force of a vast number of tiny vortices or eddies in the medium, all spinning at enormous speeds and having their axes parallel to the lines of force.[4]

Subjecting such a system to mathematical analysis, Maxwell showed that the velocity at the circumference of each vortex must be proportional to the intensity of the magnetic force and that the density of the substance of the vortices must be proportional to the capacity of the medium for magnetic induction.[5] But Maxwell was in a quandary:

> I have found great difficulty in conceiving of the existence of vortices in a medium, side by side, revolving in the same direction about parallel axes. The contiguous portions of consecutive vortices must be moving in opposite

directions; and it is difficult to understand how the motion of one part of the medium can coexist with, and even produce, an opposite motion of a part in contact with it.[6]

Maxwell's solution was ingenious. He suggested that the vortices were separated by a layer of tiny particles, each revolving on its own axis in a direction opposite to that of the vortices. These particles were supposed to be much smaller than the vortices, which were themselves conceived as being far smaller than ordinary molecules, and they were supposed to mesh with the circumference of the vortices, much as 'idle wheels' mesh with a train of mechanical gears. The axes of these 'idle wheels' were supposed not to be fixed in relation to those of the vortices, but they were supposed to be capable of a lateral or translational movement whereby the vortices in any given layer might be permitted to possess an instantaneous rate of rotation differing from those in the layer above or below.

The ingenuity of Maxwell's hypothesis paid a remarkable dividend, for on mathematical analysis of the motions of the system it emerged that a lateral movement of the particles was the equivalent of an electric current. Furthermore, in permitting the vortices in successive layers to revolve at different rates, the hypothesis suggested that the phenomenon of induced currents was part of the process of communicating the rotary velocity of the vortices from one part of the field to another.

Maxwell did not for one moment suppose that his system of vortices and particles represented the real state of affairs. Indeed, he declared categorically:

> We have now shown in what way electromagnetic phenomena may be imitated by an imaginary system of molecular vortices . . . I do not bring it forward as a mode of connexion existing in nature, or even as that which I would willingly assent to as an electrical hypothesis. It is, however, a mode of connexion which is mechanically conceivable, and easily investigated, and it serves to bring out the actual mechanical connexions between the known electromagnetic phenomena; so that I venture to say that anyone who understands the provisional and temporary character of this hypothesis will find himself rather helped than hindered by it in his search after the true interpretation of the phenomena.[7]

In the last of his series of papers in the *Philosophical Magazine*,[8] Maxwell developed the consequences which seemed to derive from the construction of his hypothetical medium. He showed that it was necessary to suppose that the substance of his vortices possessed the quality of *elasticity,* 'similar in kind though different in degree to that observed in solid bodies'. Developing the theory still further, he showed that the velocity of propagation of transverse vibrations through such an elastic medium must be equal to the ratio of the units of magnetic force to the units of electric current and, using data from the experiments of Messrs Weber and Kohlrausch, Maxwell declared that the velocity of propagation would be 193,088 miles per second.

At this date (1862), the undulatory theory of light was well established. This theory had required the assumption of elasticity in the 'luminiferous medium' in order to account for the observed transmission of transverse vibrations. Remarking, however, that the velocity of light as determined by Fizeau was 195,647 miles per second, Maxwell declared:

> The velocity of transverse undulations in our hypothetical medium, calculated from the electromagnetic experiments of Messrs Kohlrausch and Weber, agrees so exactly with the velocity of light calculated from the optical experiments of M. Fizeau, that we can scarcely avoid the inference that *light consists in the transverse undulations of the same medium which is the cause of electric and magnetic phenomena.*[9]

Maxwell's achievement in thus linking the phenomena of light with those of electricity and magnetism is all the more remarkable when one considers the framework by means of which the theory had been developed - the complicated and hypothetical structure of the medium with its molecular vortices, its connecting 'particles' of electricity and its highly complicated dynamics. But, as Maxwell said, it was a system which 'seemed mechanically conceivable and easily investigated' and it certainly served to explain virtually all the known phenomena of electricity and magnetism.

It was not, however, the only theory which aimed at accounting for the observed phenomena and for the forces between bodies situated at a distance from each other. In particular, there was the comprehensive theory developed by Weber and Neumann but although this theory accounted in a remarkable way for all the phenomena of static electricity, magnetic and electrical attraction, the induction of currents, and diamagnetism, it had one major defect. Its authors had found it necessary to assume that the forces between electrical particles depended on the relative velocity as well as upon their distance apart and, in Maxwell's view, the mechanical difficulties involved in such an assumption were such as to render the theory unacceptable as an ultimate solution. He wrote:

> I have therefore proposed to seek an explanation of the facts in another direction by supposing them to be produced by actions which go on in the surrounding medium . . . The theory I propose may therefore be called a theory of the *Electro-magnetic Field* because it has to do with the space in the neighbourhood of the electric or magnetic bodies, and it may be called a *Dynamical Theory* because it assumes that in that space there is matter in motion, by which the observed electro-magnetic phenomena are produced.[10]

The above quotation is taken from the introductory paragraphs of Maxwell's paper, *A Dynamical Theory of the Electro-magnetic Field*, which he read to the Royal Society on 8 December 1864. This paper has rightly been regarded as Maxwell's greatest contribution to electrical science and it was to have such consequences for the art of electrical communication that a few further quotations from the introduction seem permissible.

> The electro-magnetic field is that part of space which contains and surrounds bodies in electric or magnetic conditions.

It may well be filled with any kind of matter, or we may endeavour to render it empty of all gross matter, as in the case of Geissler's tubes and other so-called vacua.

There is always, however, enough of matter left to receive and transmit the undulations of light and heat, and it is because the transmission of these radiations is not greatly altered when transparent bodies of measurable density are substituted for the so-called vacuum, that we are obliged to admit that the undulations are those of an aethereal substance, and not of gross matter, the presence of which merely modifies in some way the motion of the ether.

We have therefore some reason to believe, from the phenomena of light and heat, that there is an aethereal medium filling space and permeating bodies, capable of being set in motion and of transmitting that motion from one part to another, and of communicating that motion to gross matter so as to heat it and affect it in various ways.

Now the energy communicated to the body in heating it must have formerly existed in the moving medium, for the undulations had left the source of heat before they reached the body, and during that time the energy must have been half in the form of motion of the medium and half in the form of elastic resilience.

The propagation of undulations consists in the continual transformation of one of these forms of energy into the other alternately, at any instant the amount of energy in the whole medium is equally divided, so that half is energy of motion and half is elastic resilience.

The succeeding sections of this epoch-making paper are highly mathematical and in the third section Maxwell developed the 'general equations of the electromagnetic field', twenty of them in all, involving no less than twenty variables. Unfortunately, much of Maxwell's working involved partial differential equations of a high order and his free employment of Hamilton's quaternionic calculus rendered the mathematics beyond the comprehension of the vast majority of the physicists and engineers of the time. Even for those who were able to understand his mathematics there were difficulties: his theory gave no explanation of reflection and refraction and an even greater stumbling block at the time was the principle, fundamental to his theory, that currents can only flow in closed circuits. According to the contemporary physicists, a circuit containing a condenser - or, as we should now call it, a 'capacitor' - was not a closed circuit and, it was believed, a current charging the condenser terminated at the plates of the condenser. Maxwell, on the other hand, insisted that the real seat of the process lay in the dielectric, that the charging current was really a 'displacement current' in the dielectric and that since the displacement current was merely a continuation of the charging current, it might still be regarded as flowing in a continuous closed circuit.

These difficulties, and especially the need to accept the existence of displacement currents in free space, seemed to render Maxwell's theories highly controversial to the great majority of his most distinguished contemporaries. Although Helmholtz ultimately accepted it, he did so only after the lapse of many years and, in spite of the work of Hertz in 1887-8, Kelvin never seemed completely to believe in it even to the end of his life in 1907. Today, the name of Clerk Maxwell is everywhere respected and his work is regarded as the very foundation of electromagnetic theory, but to the world of the 1860s and 1870s, his name was almost unknown outside a narrow academic and scientific circle.

Maxwell has, not infrequently, been credited with having 'predicted the existence of radio waves' but, in common with other similarly loose statements, this is a half-truth and its implication inaccurate. Perhaps it is important, therefore, in the present context to state as briefly as possible the main purpose and conclusions of his famous paper on the electrodynamic field.

The main purpose of the paper was to investigate the characteristics of the medium surrounding electric and magnetic bodies and to identify this medium, if possible, with that through which the propagation of light was assumed to take place. In the course of his task, Maxwell made a very thorough investigation of the fields surrounding electrified and magnetic bodies during which he derived the 'general equations of the electromagnetic field'. Then, in contrast to the earlier theories which assumed that energy resided - in the form of potential energy - in electrified bodies, conducting circuits or magnets, Maxwell showed that energy was stored in the electromagnetic field - the space surrounding the electrified or magnetic bodies.

Maxwell then proceeded to examine whether the properties which constituted an electromagnetic field, derived from electromagnetic phenomena alone, were sufficient to explain the propagation of light in the form of transverse waves. He was able to show that such propagation would take place with a velocity determined by the number of electrostatic units in one electromagnetic unit of electricity. By the best determination of this ratio then available, the velocity of propagation of electromagnetic waves could be stated as 310,740,000 metres per second. This coincided so closely with the measured velocity of light that Maxwell was able to write:

> The agreement of the results seems to show that light and magnetism are affections of the same substance, and that light is an electromagnetic disturbance propagated through the field according to electromagnetic laws.

For some years following the publication of *A Dynamical Theory of the Electro-magnetic Field* in 1864 no substantial progress was made. The theory had not been widely acclaimed and Maxwell had retired to his estate at Glenlair, but in 1871 he returned to Cambridge as Professor of Experimental Physics and, two years later, he published his *Treatise on Electricity and Magnetism*. The famous work embraced almost every branch of electricity and magnetism but, as Sir Edmund Whittaker has pointed out, it seems to have been written from a single view-point - that of Faraday - and the *Treatise* was less successful if considered as an exposition of its author's own views. The doctrines peculiar to Maxwell - the existence of displacement currents and of electromagnetic vibrations identical with those of light - are not dealt with until the second half of the second volume and the account then given is not greatly different from that in the original paper to the Royal Society in 1864.

Maxwell died in November 1879 at the early age of forty-eight, ten years too early to witness the brilliant confirmation of his theories at the hands of Heinrich Hertz.

David Edward Hughes (1831-1900)

Following, so far as practicable, a chronological sequence, a reference must now be made to the confirmation of Maxwell's theories which was so nearly provided within a few weeks of his death by the work of David Edward Hughes, the inventor of the microphone and of the printing telegraph which bears his name. While carrying out experiments on his induction balance during the autumn of 1879, Hughes was led, by the accident of a bad contact in the circuit of the Bell telephone he was using as a detector, to investigate certain strange effects which seemed to arise from what, in the language of the day, was known as the 'extra current', which was observed in an inductive circuit when the current was interrupted. It was not, of course, generally realized in those days that the so-called 'extra-current' was an oscillatory transient of high frequency.

Hughes carried out many experiments during the closing months of 1879 and found that the inclusion of one of his experimental microphones in the circuit enabled him to detect the presence of an 'extra current', even when his detecting circuit was removed to a considerable distance. In later years it became evident that Hughes had, in fact, discovered a high-frequency rectifier and that he was really detecting electromagnetic oscillations radiated from the primary circuit. He was able to detect the sounds of the 'extra current' at quite considerable distances. He wrote in 1890:

> After trying unsuccessfully all distances allowed in my residence in Portland Street, my usual method was to put the transmitter in operation and walk up and down Great Portland Street with the receiver in my hand, with the telephone to my ear. The sounds seemed to slightly increase for a distance of 60 yards and then gradually diminish, until at 500 yards I could no longer with certainty hear the transmitted signals.

Hughes demonstrated his experiments to a number of distinguished scientists of the day, including William Preece and Sir William Crookes but it was the demonstration to Mr William Spottiswoode, then the President of the Royal Society, Professor Huxley, and Sir George Stokes on 20 February 1880 which was to have such unfortunate consequences for Hughes and for the history of our subject. Towards the end of a lengthy demonstration of his experiments at which they had initially been impressed with his results, Professor Stokes and his colleagues expressed the view that the results could be explained by simple induction and they refused to accept Hughes' opinion that they were due to 'conduction through the air'.

DAVID EDWARD HUGHES
1831-1900

Figure 4   David Edward Hughes

It is easy to be wise with the benefit of hindsight and there can now be no doubt that Hughes was, in fact, generating and detecting electromagnetic waves of high frequency. It is equally certain that Hughes himself, whose scientific knowledge was very limited, had no real understanding of the phenomena he was demonstrating. In a letter to J. J. Fahie[11] written twenty years later, in which he complains bitterly of the discouraging treatment he had received from Stokes, Spottiswoode and Huxley, Hughes refers repeatedly to the 'transmitter' and the 'receiver' and to 'aerial electric waves', but this was not his terminology in 1879-80. If only he had referred to the notes[12] he made on the very evening of his unhappy demonstration to Stokes and his colleagues, he would have recalled that his own explanation of the phenomena was based on *conduction and nothing else*.

Stokes and his companions were equally wrong in maintaining that the experiments could be explained by simple induction but they will have realized that Hughes was a man of very limited scientific knowledge, (he was a professor of music, not of physics), and his apparatus was crude and home-made. They were themselves not specialists in the field of electromagnetic physics and, while it was most unfortunate that the weight of their opinion was to have such a discouraging effect on Hughes that he refused to publish any contemporary account of his experiments, they might perhaps be partially excused on the grounds that this was one of those very rare instances in which a man with little scientific background happened to stumble across a scientific principle of great significance. The fact remains, however, that had they been a little less categoric and a little less discouraging upon a matter in which none of them were authorities, the early history of radio communication might have taken a very different line.

George Francis FitzGerald (1851-1901)

Maxwell's electromagnetic theory was, as we have seen, a theory only lacking - until the work of Hertz in 1887-8 - any physical confirmation. Its complexity and, in particular, its mathematical presentation, rendered it difficult to comprehend, and it is safe to say that at the time of Maxwell's death in 1879 barely a handful of mathematical physicists unreservedly accepted its implications.   Even Helmholtz and Thomson (Lord Kelvin) were among the sceptics, though the former ultimately came to accept it.

The most persistent and influential advocate of Maxwell's view was unquestionably Professor George Francis FitzGerald of Trinity College, Dublin, one of the outstanding physicists of his time and one who, by common consent, had attained a unique position as a judge and critic of the work of others.

FitzGerald's first 'declaration of faith' in Maxwell's theory was contained in his paper, *On the Electro-Magnetic Theory of the Reflection and Refraction of Light*[13], which he addressed to the Royal Society in January 1879. Maxwell had dealt with the propagation of electromagnetic waves through uniform media but he had left untouched the questions of reflection and refraction. FitzGerald succeeded in showing that the solutions to these questions also were completely embraced within Maxwell's system of equations and, further, he showed that Maxwell's theory led to laws of reflection and refraction identical with those earlier deduced by MacCullagh from the undulatory theory of optics.

FitzGerald's support was a valuable contribution but, before Maxwell's theory could be accepted as established fact, it was necessary to demonstrate *either* that ordinary light could be produced by purely electromagnetic means *or* that waves generated by an electromagnetic process possessed all the properties and physical characteristics normally associated with visible light. The difficulties, however, were formidable.

As FitzGerald well knew, the frequency of the waves of visible light was of the order of six hundred million million (6×10^{14}) cycles per second, corresponding to a wavelength of only about 1/2,000 mm, and with the techniques then known, it was clearly impossible to conceive of any means of originating such waves by electromagnetic means. At the other end of the scale were the low frequencies, and while it would have been possible to generate low frequency waves, the corresponding wavelengths would have been impossibly long for

Figure 5   George Francis FitzGerald

experimental purposes, apart from which, as FitzGerald pointed out, the radiated energy would have been negligible. It was thus clear to FitzGerald that the possibility of carrying out experiments on electromagnetic wave radiation all depended upon the ability to generate alternating currents having a frequency of the order of 10-100 million cycles per second. In his paper,[14] *On the Possibility of Originating Wave Disturbances in the Ether by means of Electric Forces*, which he addressed to the Royal Dublin Society in 1882, he declared: 'It might, however, be possible to obtain sufficiently rapid alternating currents by discharging condensers through circuits of small resistance'. Addressing the British Association in the following year, he went even further and stated categorically:

> By utilizing the alternating currents when an accumulator (i.e. a condenser or Leyden jar) is discharged through a small resistance, it would be possible to produce waves of as little as 10 metres wavelength, or even less.[15]

Knowing how suitable waves might be produced was one thing; the ability to detect them was quite another and, at the time, there was no known method by which the physical existence of such waves could be detected. As we might remark casually today, 'the air is full of radio signals', but without the appropriate means for receiving them, they 'pass over our heads' and we remain completely unconscious of their existence. So it was with FitzGerald; on theoretical grounds he was convinced of the physical possibility of their existence and he had proposed a method by which they might be generated - but actually to detect them, and thus to demonstrate that they really did exist, was another matter altogether.

Another three years were to pass before Heinrich Hertz was to observe - almost by accident - the tiny sparks in a resonant circuit which gave him the clue to the means whereby the waves could be detected. As we shall see, it was this observation in 1886 which was to open the way for his epoch-making experiments and, in 1888, for his physical confirmation of Maxwell's theoretical predictions. But, as Sir Oliver Lodge was to comment later,[16] it seems beyond doubt that it was FitzGerald who recognized even more vividly than Hertz himself, the full import of Hertz' discovery. Indeed, the almost immediate recognition and appreciation which was accorded to Hertz' experiments was very largely due to the quick and clear vision displayed by FitzGerald and, in particular, to the attention which he so generously focused upon those experiments in his Presidential Address[17] to Section A of the British Association at its meeting in Bath during September 1888. It is not often given to a President to announce in his address some epoch-making scientific discovery but such an opportunity fell to FitzGerald and few could more have deserved the privilege or put it to better use.

Although it is true that FitzGerald made no separate discovery or invention which could be said to have constituted a recognizable stage in pre-radio history, the unqualified endorsement which he gave to Maxwell's theory, the very full support he gave to Oliver Lodge and, as we shall see later, the encouragement he subsequently gave to Heinrich Hertz himself, more than justify the inclusion of his name among the pioneers.

References

1. William Thomson, 'On a Mechanical Representation of Electric, Magnetic and Galvanic Forces', *Cambridge and Dublin Mathematical Journal*, 1847, **2**, pp.61-64.
2. J. Clerk Maxwell, 'On Faraday's Lines of Force', *Transactions of the Cambridge Philosophical Society*, 1856, **10**, pp.27-83.
3. J. Clerk Maxwell, 'On Physical Lines of Force, Part I: The Theory of Molecular Vortices applied to Magnetic Phenomena', *Phil. Mag.* (4th series) 1861, **21**, pp.161-175; 'On Physical Lines of Force, Part II: The Theory of Molecular Vortices applied to Electric Currents', *Phil. Mag.* (4th series), 1861, **21**, pp.281-291, 338-348.
4. Maxwell, 'On Physical Lines of Force, Part I', p.165.
5. Maxwell, 'On Physical Lines of Force, Part II', p.262.

6. *ibid.*, p.283.
7. *ibid.*, p.346.
8. J. Clerk Maxwell, 'On Physical Lines of Force, Part III: The Theory of Molecular Vortices applied to Statical Electricity', *Phil. Mag.* (4th series), 1862, **23**, pp.12-24.
9. *ibid.*, p.22.
10. J. Clerk Maxwell, 'A Dynamical Theory of the Electromagnetic Field', *Phil, Trans.*, 1865, **155**, pp.459-512. (A reprint was published in *The Scientific Papers of James Clerk Maxwell*, **1**, pp.526-597. For a non-mathematical abstract of this famous paper, see *Phil. Mag.* (4th series), 1865, **29**, pp.152-157, or *Proceedings of the Royal Society* (hereafter *Proc.Roy.Soc.*), 1864, **13**, pp.531-536.
11. J. J. Fahie, *A History of Wireless,* William Blackwood & Sons, 1900.
12. D. E. Hughes, Manuscript Note Books, 1878-1886, in British Library. MSS Nos. 40161/3.
13. G. F. FitzGerald, 'On the Electro-Magnetic Theory of the Reflection and Refraction of Light'. *Phil. Trans.*, 1880, **171**, Art.XIX., pp.691-711. For an abstract see *Proc. Roy. Soc.*, 1878-9, **27**, pp.236-238. For reprints of both see J. Larmor, *The Scientific Writings of the late George Francis FitzGerald*, Longmans Green, 1902.
14. G. F. FitzGerald, 'On the Possibility of Originating Wave Disturbances in the Ether by means of Electric Forces', *Scientific Transactions of the Royal Dublin Society*, Series 2, 1877-83, **1**, pp.133-134, 173-176, 325-326. For reprints see Larmor, *op.cit.*
15. British Association Report, Southport Meeting 1883, Transactions of Section A, No.6, p.405. Reprinted in Larmor, *op.cit.*
16. O. J. Lodge, Obituary article, 'G. F. FitzGerald'. *Proc. Roy. Soc.*, 1898-1905, **75**, pp.152-160. Reprinted in Larmor, *op.cit.*
17. G. F. FitzGerald, Presidential Address to the Mathematical and Physical Section of the British Association for the Advancement of Science, 1886. See also Larmor, *op.cit.*

Chapter 4

Hertz

Hertz and Lodge

At this stage in the early history of radio, around 1880, the historian finds himself faced with a difficult problem. Hitherto, the story has developed in a comparatively orderly manner, the mantle of one pioneer descending almost as of right upon the shoulders of his automatic successor. First there was Oersted whose observation of the effect of an electric current flowing in a wire just above a compass needle was subsequently to be taken up by Faraday by whom, over a period of many years, the concept of lines of force was gradually developed to form the background to what we would now call field theory.

Faraday, as we have seen, was not an academically trained physicist. His appointment as Davy's assistant at the Royal Institution placed him in a highly privileged position but, lacking any formal education, a majority of his contemporary scientists were reluctant to accept him as a properly qualified 'member of the club' and his theories were treated with more than their fair share of scepticism. Indeed, it was not until his concept of lines of force was taken up by the young Cambridge wrangler, Clerk Maxwell, and expressed in a conventional mathematical form that a cloak of respectability began to descend on the bare shoulders of Faraday's ideas.

In Maxwell's hands, Faraday's concept was destined to emerge in 1864 as the *Dynamical Theory of the Electro-magnetic Field* and to lead to Maxwell's affirmation that the phenomenon of light was, in fact, a manifestation of electromagnetic wave propagation. But for more than twenty years, Maxwell's theory was to remain a theory only, accepted by very few of the mathematical physicists of the day. With the approach of the 1880s, however, we begin to see real progress towards the discovery of a demonstrable proof of the theory and, as a consequence, the final rejection of the theory of action at a distance.

While the story has followed a logical and sequential progress so far it now branches into two roughly parallel channels, in each of which there are major contributions. With each channel developing simultaneously and almost independently of the other, it is not easy to decide to which to give priority.

The two channels were respectively those pioneered by Heinrich Hertz and Oliver Lodge. The ultimate aim of both was identical - to provide a demonstrable proof of Maxwell's theory that electrical and magnetic phenomena are propagated as transverse waves through space with the velocity and characteristics of visible light. At the time, the task was one of purely academic and philosophical interest, for neither Hertz nor Lodge had the slightest idea that there might be any practical outcome to their work. Furthest of all from their thoughts would have been any idea that they might be laying the foundation for a system of practical radio communication!

To Hertz clearly belongs the honour of being the first actually to demonstrate the reality of electromagnetic waves in the air; to Lodge, whose interest in electromagnetic waves was aroused earlier than that of the rather younger Hertz, unquestionably belongs the credit of playing a vital role in the introduction and early development of practical radio communication.

Hertz' work in this field was virtually complete by 1890 - and when he died so prematurely on New Year's Day 1894, it was almost inevitable that Lodge should be invited to deliver a Memorial Lecture at the Royal Institution.

Throughout the 1880s both Hertz and Lodge were intent on establishing the truth of Maxwell's theory. The honour of being 'first past the post' clearly belongs to Hertz but, had he not succeeded, it is likely that Lodge would have done so within a short space of time. With Hertz' untimely death, however, it fell to Lodge to make the greater contribution towards the development of practical radio communication.

Among the other leading scientists of the day, Lodge stands pre-eminent as the one with by far the deepest appreciation of the physical background and the practical realities of electromagnetic wave generation. No one had a better understanding of the mathematical basis underlying oscillating currents of high frequency or, for example, of the principles and need for resonance between radio-frequency circuits.

On the grounds that Hertz was the one first actually to demonstrate the reality of electromagnetic waves through the air, while Lodge's interest extended well into the era of practical radio communication during the early years of the present century, let us deal first with Hertz' career and subsequently with that of Lodge, while remembering that until 1894 the careers of the two men were substantially contemporary and followed closely similar lines.

Heinrich Hertz (1857-1894)

As we saw earlier, David Hughes had come very near to a practical demonstration of the truth of Maxwell's theories in 1880 and, had he done so, he would have gained the honour of being the discoverer of electromagnetic wave propagation. But whatever view one may take of his unfortunate interview with the 'three wise men', the fact remains that Hughes himself was not a scientist and his observations were entirely accidental. His work was largely fortuitous and he

Figure 6   Heinrich Rudolf Hertz

had very little understanding of the phenomena with which he was dealing. In contrast, the work of Heinrich Hertz was that of a trained and brilliant physicist who had the understanding and the ability to appreciate the meaning of a chance observation and to develop logically, step by step, the magnificent series of experiments which demonstrated all the properties of electromagnetic radiation and conclusively established its identity with the properties of light.

Education

Heinrich Rudolf Hertz was born on the 22 February 1857, the eldest son of Dr G. F. Hertz, a prominent lawyer who subsequently became an appeal court judge and, later, a leading Senator of the Hanseatic city of Hamburg. His mother was the daughter of a physician, Dr Johann Pfefferkorn of Frankfurt-am-Main. Three more sons followed Heinrich, Gustav only a year later, Rudi in 1861 and Otto in 1867; finally a daughter, Melanie, was born in 1873 to complete the family.

From a charming memoir composed by his mother some years after his death,[1] it is clear that her eldest son showed outstanding promise even from his earliest days. Both as a child and throughout his later years, his memory was phenomenal and he possessed an almost insatiable appetite for learning. Not only was his aptitude for mathematics exceptional but he balanced this with a determination to excel at languages. The Greek and Latin classics held no terrors for him and, even in adult life, he retained a passion for Homer whose verses, as well as those of Dante, he could recite at length.

From an early age, young Heins - as he was known in the family - also showed exceptional ability with his hands. Adept in the use of tools of all kinds, it seemed likely that he would eventually take up some form of practical occupation. Mechanical engineering seemed a possibility but, such are the whims of chance and such the uncertainties which confront the teenager, that a trivial event at the age of seventeen almost succeeded in diverting him into quite another direction. On a street-vendor's barrow, he chanced one day to pick up an Arabic grammar and, with his interest in languages already aroused, he began to study it with enthusiasm. His father, thinking to make the task easier, advised him to consult a specialist, Professor Redslob, who gladly accepted him as a student. Before many weeks had passed, Professor Redslob called on Dr Hertz and implored him to encourage the young man to become an orientalist. Never before, he said, had he come across a student with such an aptitude for oriental languages. Fortunately, the immediate need to concentrate on his matriculation examination provided a diversion and by the time the exams were over the temptation had passed. From his mother's account, however, it was a 'near thing' and, had it succeeded, it is clear that not only would the life of Heinrich himself have developed along very different lines but the whole history of radio would have taken a different form.

Not long after young Heinrich left the private school at which he had been a pupil for almost nine years, he recorded his thoughts on the teaching and discipline which he had there experienced. The contemporary words of a seventeen-year-old provide an interesting description of the school's regime, and provide a fascinating commentary on the young man himself.

> All of us, or at least the better scholars amongst us, were unusually fond of Dr Lange's school, despite the hard work and the great strictness. For we were ruled strictly; detentions, impositions, bad marks for lack of neatness or ill-behaviour rained down upon us; but what particularly sweetened the strictness and profusion of work for us was, in my opinion, the lively spirit of competition that was kept alert in us and the conscientiousness of the teachers who never let merit go unrewarded nor error unpunished.

Now began two years of private study at home. Every morning from seven to eight o'clock (in winter from eight to nine) he had a lesson: Greek and Latin twice a week from Dr. Kostlin, mathematics twice a week with Herr Schottke and on Sunday he attended the *Gewerbeschule* [technical school] from nine to twelve noon. At the end of the daily lesson and on his mother's insistence he went to a physical exercise class but by nine o'clock he was back at his work at home where he remained completely undisturbed until lunch time. Half an hour's play with his youngest brother was followed by more study until dinner time at five o'clock.

After two years, at Easter 1874, he was accepted into the top class of the upper school of the Hamburg Gymnasium, a school in some ways comparable with a modern 'Sixth Form College'. A year later he left with a certificate which qualified him for entry to a university.

Apprentice engineer

The time had now come for Heinrich to decide on his future career. Unlike Oliver Lodge who, from his early teens, had been determined to pursue a career in the natural sciences, young Hertz had made no early choice and his studies had tended to follow a middle course, divided almost equally between the classics and mathematics. He had acquired considerable manual dexterity and much of his spare time was occupied in the construction of mechanical devices. Indeed, it is not impossible that his proven manual skills influenced him in making his first choice of a career towards civil engineering rather than towards the more esoteric field of pure science, although it is evident from an essay he wrote at the age of eighteen that the possibilities of a scientific career had not been dismissed from his thoughts:

> I intend, if I succeed in passing the matriculation examination, to go to Frankfurt-am-Main and work for a year under a Prussian architect, as I would be ultimately required to do for the state licensing examination for professional engineers. Only if I were to prove unsuited for this profession or if my interest in the natural sciences were to increase further, would I devote myself to pure science. May God grant that I choose whatever I am best suited for.

The year which Heinrich spent as an apprentice in the city engineer's office at Frankfurt provided spare time to read the many books he was able to borrow from the municipal library. His choice of literature is interesting, the Greek classics predominating, but one cannot help wondering just how many young men in a similar position today would happily choose to spend a large proportion of their spare time immersed in the works of Aeschylus, Euripides and Plato! Young Heinrich's reading list was not entirely confined to the classics. One of the books mentioned repeatedly in his diary is Wüllner's *Physics* which he studied assiduously. An entry for the week ending 5 November 1875 reads: 'I am reading a great deal in Wüllner's *Physics* and am turning once again towards natural sciences. Even though I am not allowing my old desire to study natural sciences to rise uppermost again, yet I should very much like to do my military service next year in Hamburg so that I could study natural science in my spare time which I could do better there than elsewhere.'

Heinrich's difficulty in choosing a career was related to the innate modesty which was one of the outstanding features of his character throughout his whole life. Highly talented young people not infrequently seriously underestimate their own potential. Unable to see into the future and lacking the experience to make a considered judgement, it is likely that young Hertz tended to regard a scientific career as an attainment altogether beyond his limits. Like so many youngsters, he could appreciate the difficulties in plotting such a path but he had a far too slight regard for his own capabilities.

His year in the municipal engineering office at Frankfurt was up in April 1876 but although he was still unable to make up his mind, his level of self-confidence had been raised by attending a series of science lectures at the Frankfurt Physics Club. His year of military training was looming, however, and he resolved to spend the few months which intervened studying higher mathematics - which he found 'incredibly interesting' - at the Technical High School in Dresden.

Hertz would never have complained that his year of military service was a waste of time: writing to his parents he commented 'the drill itself is generally exhilarating and we are always in good spirit. Furthermore, it is obvious that hunger is the best cook and we sleep splendidly.' Later he wrote, 'The service will have one lasting effect, for laziness is driven out of one's system and one soon realizes how much one can do when one must. I often think that if I had been prepared to work away at my lathe and plane with quite the same zeal and recklessness about my fingers with which I must now practice rifle-drill, I should have accomplished much more in the same time. Or if I had always obeyed my better judgement as quickly and exactly as I now have to obey the orders of an officer or N.C.O., orders which may seem quite arbitrary to me, I should have spared myself many worries.'

Engineering or science?

Hertz completed his year of military service at the end of September 1877 still undecided as to his future career but wavering towards civil engineering. With this in mind he resolved to take the degree course in that subject at the College of Technology, the *Polytechnicum*, in Munich. But, as he read over the syllabus of the classes he would attend, he realized more clearly than ever before the real nature of the work which, as an engineer, would occupy him in the immediate future and, in all probability, for the rest of his life. He realized that he had little interest in the surveying and designing, the construction and specifying which occupy a civil engineer, and that if he chose to follow such a course it would only be because it might provide a shorter road towards financial independence from his parents than would a career in natural sciences for which, in the depths of his heart, he really yearned.

It is seldom that one has the privilege of being privy to the innermost thoughts of a young man as he contemplates his choices at such a crucial stage of his career and as he realizes that he has reached the cross-roads at which he must make an irrevocable decision. Happily, Heinrich's letters to his parents have survived[2] and among them that which he wrote to his father on 1 November 1877, seeking his approval for the switch from engineering to natural sciences upon which he had at last decided:

It is a shameful confession for me to make, but out it must come: I should like to change horses now, at the last moment, and switch to the natural sciences. In this semester I arrive at the crossroads where I must either devote myself to science completely or take leave of science for good and give up all superfluous dallying with the subject if I am not to neglect my proper studies and become merely a mediocre engineer. When I recently came to realize the situation beyond all possible doubt, my first thought was to renounce all my dithering with mathematics and science and devote myself to engineering; but then I realized that this was something I could not bring myself to do, that I had so far only occupied myself with these subjects and looked forward to them alone . . .

I cannot understand why I did not realize this before now for even in coming to Munich, my mind was set on studying mathematics and the natural sciences and I had no thought at all about surveying, building construction, builders' materials, etc., which were supposed to be my main subjects . . . I now realize that what are called the engineering sciences is not enough to satisfy me . . . and I now see that this is the reason I am always moving from one school to another . . . but I am sure of one thing, in the natural sciences I shall not yearn to be back in engineering science, whereas if I became an engineer, I would always be longing for science . . . looking back, I realize that I have had ten times the encouragement to study natural sciences than to become an engineer, and it is doubtful whether as an engineer I would in the end have any advantage over others simply because of my possibly somewhat better grounding in mathematics as it seems to be that, at least for the first ten years in the profession, much more depends in the end on experience, common sense and a knowledge of data and formulae which, since they are incidental, do not interest me . . . I have come to the conclusion that while it might be of some tangible advantage to become an engineer, it would involve a kind of self-denial and renunciation which I should not force upon myself if external reasons do not compel me to do so . . .

And so I am asking, dear Papa, not so much for your advice as your decision . . . if you tell me to study natural science, I shall take it as a great gift from you and I shall do so with all my heart. I do believe that this will be your decision, partly because you have never placed an obstacle in my path and partly because you have sometimes seemed to prefer to see me studying science. But if you consider it best for me to continue along the road on which I have set out but in which I no longer believe, I shall do that, too, and do it without reservation for I have had enough of doubt and hesitation . . .

If I carry on as I have been doing, I shall never advance. It is now or never. I don't want to lose any more time and I regret the time I have already lost. Please give me your answer soon. Most of the courses begin next Monday and there is no time to waste. If your decision turns out as I hope, I will withdraw my name from the Polytechnic rolls and see the physics professor at the University to seek his advice as to how I should apportion my time and exactly which programme I should pursue . . .

His father's reply has not survived but it is evident from Heinrich's next letter that it gave him the approval he had so earnestly sought. Warmly thanking his father, he tells him, 'I enrolled right away at the University and went to see the physics professor, Dr von Jolly, to ask his advice as to the courses I should take.' He settled on advanced mathematics and mechanics, experimental physics and experimental chemistry - 'and I rejoice in my new activities and in the thought that my studies will not now come to an end in a few years, but should continue all my life. Naturally, one is never quite content, and so I am burning with impatience to reach the frontier of what is already known and to go on exploring into unknown territory . . .'

Avid for knowledge, his letters reveal that he was never happier than when deeply engrossed in the study of some abstruse branch of higher mathematics, but in Heinrich's philosophy there were no short cuts. He seems clearly to have realized that if he was eventually to succeed in his object of taking his place among the leading physicists of the day, he would first have to make himself a complete master of existing knowledge and he was prepared to devote himself with unrelenting enthusiasm to the hard study by which alone he could achieve his object.

Although it was his aim to reach the very 'frontiers of science', he seems to have been doubtful as to whether there would be anything left to discover when he got there. In one letter written to his parents in February 1878 he expresses his regret that he had not lived in an earlier age. In the light of our knowledge of his own momentous discoveries only ten years later, it is almost amusing to read his words: 'I do not think that it will be possible to discover anything nowadays that would lead us to revise our entire outlook as radically as was possible in the days when the telescope and microscope were new . . .'

Hertz spent only a year at Munich but it was a vital year in his career, a year which saw his final conversion from a man of practice to one of learning and this pursuit of learning *per se* encouraged him in the autumn of 1878 to transfer to the University of Berlin where he quickly attracted the attention of the celebrated Professor Hermann von Helmholtz.

Academic career

Helmholtz drew the attention of the young student to a problem for the solution of which a prize was offered by the Philosophical Faculty of the University. The problem was an old one and involved the theory that electricity was a fluid which physically moved through any conductor conveying it. But if electricity was a fluid, it must possess mass, even if that mass was of very small dimensions; and if electricity possessed mass, it must also possess the properties of inertia and kinetic energy when in motion. This kinetic energy, it was argued, should become manifest when the current flowing in any circuit was suddenly interrupted.

It was known that the sudden making or breaking of a circuit was accompanied by certain phenomena which, in the terminology of the period, were known as 'extra-currents', and it was assumed that these so-called 'extra-currents' were connected with the supposed mass and kinetic energy of the current. In view of the difficulties of the measurements involved, it was not expected that any of the students attempting the problem would be able to arrive at an exact estimate of the kinetic energy, and hence of the quantity of electricity in any given volume of the conductor, but it was hoped that the research involved might provide an answer to the question: 'How great can this quantity of electricity possibly be?'

It was perhaps a little unfortunate that this first research - which occupied young Hertz for most of the winter - should have had a negative result in that the more carefully he prepared the experiments and the more precisely he carried out the measurements, the more he was compelled to conclude that if all the possible sources of error could be eliminated, the quantity he sought would eventually prove to be zero. In spite of the negative outcome, the research was of tremendous importance for Hertz himself. Not only did his paper, *Experiments to determine an Upper Limit for the Kinetic Energy of an Electric Current*,[3] easily win him the prize, but it also earned him the lavish praise of the faculty and, in particular, that of von Helmholtz who had clearly recognized the exceptional abilities of his young student. When a vacancy for an assistant professor at the Physics Institute of Berlin arose, von Helmholtz offered the post to Hertz who was glad to accept the opportunity of an appointment as personal assistant to the man whom he so greatly admired. In the meantime Hertz had been awarded his Doctorate - with the rare distinction *magna cum laude*, for a dissertation *On the distribution of electricity over the surface of moving conductors*.[4]

Soon after taking up his new appointment in September 1880, his attention was drawn by von Helmholtz to the outstanding need for a proof of Maxwell's theory that electromagnetic forces were propagated through space with the same velocity as that of light and that, in fact, light itself was an electromagnetic phenomenon. An integral component of the theory asserted that an electromagnetic force would give rise to a 'displacement current' in any non-conductor subjected to an electromagnetic field. Any proof that the electrical state of an insulator had any relation with electromagnetic forces to which it might be subjected would amount, in effect, to a proof of Maxwell's theory and, accordingly, the Berlin Academy of Science had offered a prize for research 'to establish any relationship between electromagnetic forces and the dielectric polarization of insulators - that is to say, either an electromagnetic force exerted by polarizations in non-conductors, or the polarization of a non-conductor as an effect of electromagnetic induction'.

In inviting Hertz to undertake the research, von Helmholtz was, in effect, inviting him to attempt to find a proof of Maxwell's theory. It was a task for which he was ideally qualified but after giving the problem long and careful consideration, he was forced to conclude that with the facilities which were then available it would not be practicable to reach a satisfactory conclusion. A few years later, he was to write:

I reflected on the problem, and considered what results might be expected under favourable conditions using the oscillations of Leyden jars or of open induction-coils. The conclusion at which I arrived was certainly not what I had wished for; it appeared that any decided effect could scarcely be hoped for, but only an action lying just within the limits of observation. I therefore gave up the idea of working at the problem; nor am I aware that it has been attacked by anybody else. But in spite of having abandoned the solution at that time, I still felt ambitious to discover it by some other method; and my interest in everything connected with electric oscillations had become keener. It was scarcely possible that I should overlook any new form of such oscillations, in case a happy chance should bring such within my notice.[5]

Maxwell, it will be recalled, had died in 1879, a date which virtually coincided with the beginning of Hertz' career so it is unlikely that there would ever have been any communication between them. Nevertheless, by 1881 Hertz had become fully conversant with Maxwell's theory and, having been alerted by von Helmholtz, he was fully aware of the need to provide a demonstrable proof. It was thus a considerable disappointment to him to have to conclude that, in the prevailing circumstances, a demonstrable proof of the theory was out of reach. The challenge remained, however, as subsequent events were to show. Hertz remained on the look out for any advance which might render a proof possible.

There were two principle areas of difficulty which, in 1881, led Hertz to conclude that no solution to the problem was in sight. Firstly, there was the fact that no adequate means of detecting the presence of electromagnetic waves was available, and, perhaps more difficult, no means appeared to exist for the generation of electromagnetic waves of an appropriate order of frequency.

Following his analysis of the problem, Hertz had concluded that an investigation of the properties of 'Maxwellian' electromagnetic waves which could possibly be carried out within the confines of a laboratory would call for the use of waves with a wavelength of between about 50 cms and 5 metres. The trouble was that no one knew how to generate such waves, the very existence of which was still only a matter of theoretical assumption.

Hertz was up against almost exactly the same problems as were currently being confronted by Professor FitzGerald in Dublin but, as we shall see, it was to be Hertz, some four or five years later, to whom the honour would fall of finding the solution and of establishing the truth of the Maxwell theory.

By 1883 Hertz had been Professor Helmholtz' assistant for three years but, if he was to achieve promotion in the academic world, the time had come for him to move on. With some misgivings, he accepted an appointment at the University of Kiel, his doubts arising from the fact that while the post offered some prospects of promotion within two or three years, the university, being small, lacked adequate laboratories and there were none of the workshop facilities which Hertz had come to regard as essential for the construction of his experimental apparatus. However, the post offered experience in teaching and lecturing and, in so far as it left him with ample time for contemplative study and theoretical developments, he was undoubtedly wise to accept it.

An important product of the 'Kiel period' was the publication in 1884 of Hertz' paper, *On the Relations between Maxwell's Fundamental Electromagnetic Equations and the Fundamental Equations of the Opposing Electromagnetics*,[6] but to appreciate its significance, one must realize that Hertz' support for Maxwell's theory was no sudden phenomenon. It was, in fact, spread over a period of several years. As a pupil of von Helmholtz, Hertz had been indoctrinated with the continental theories of electrical phenomena which were currently accepted in Germany, and which were still primarily based upon the action at a distance theories of the eighteenth century. In considering the terms of the Berlin Academy prize in 1880, he had begun seriously to question the accepted theory and to support Maxwell's belief that electromagnetic waves were propagated through space with the same velocity as waves of light and that light itself consisted of electromagnetic waves. In his Kiel paper of 1884, Hertz subjected the Maxwell theory and those 'of the opposing electromagnetics' to rigorous mathematical analysis. While he was unable to reach a final decision as to the relative merits, he concluded by saying, 'I think we may infer without error that if the choice rests only between the usual system of electromagnetics and Maxwell's, the latter is certainly to be preferred'. He remained acutely aware that the choice could not finally be settled until someone could, by some practical demonstration, show that waves generated by electromagnetic means possessed all the physical qualities of light and differed only in frequency.

After less than two years at Kiel and at the age of barely twenty-eight, Hertz received the offers of no less than three professorships - at Greifswald, at Kiel itself and at Karlsruhe. Attracted particularly by the laboratories and the facilities they provided for the research programme he hoped to pursue, he accepted the post at the Technical High School at Karlsruhe where, as he was soon to discover, the laboratories were not the only 'attraction'. In July 1886, the year following his appointment to Karlsruhe, he married Elizabeth Doll, the daughter of Dr Max Doll, the Lecturer in Geometry at the Technical High School.

The need for a proof of Maxwell's theory was never far from Hertz' mind and he was unlikely to miss a clue if one should arise fortuitously. Among the collection of physical apparatus at Karlsruhe he had found an old pair of 'Knockenhauer Spirals' - short fat coils of wire embedded in spiral tracks cut into the surface of circular wooden discs - and, while experimenting with these coils during the autumn of 1886, he observed a small spark passing between the terminals of one of the coils whenever a Leyden jar was discharged through the other. It may have been tiny but that spark set alight the train of research which, during the next four years, was to establish Hertz among the leading scientists of the nineteenth century.

Figure 7   Knockenhauer Spirals

High frequency oscillations

It is easy for us to recognize the phenomenon as one arising from simple resonance between one circuit and another, but to Hertz, as it would have been to any contemporary, the occurrence of these sparks between the ends of an unconnected coil was a matter calling for further enquiry. In Hertz' own words:

> I had been surprised to find that it was not necessary to discharge large batteries through one of these spirals in order to obtain sparks in the other; that small Leyden jars amply sufficed for this purpose and even the discharge of a small induction coil would do, provided it had to spring across a spark gap. At first I thought the electrical disturbances would be too turbulent and irregular to be of any further use; but when I discovered a neutral point in the middle of a side-conductor, and indications, therefore of a clear and orderly phenomenon, I felt convinced that the problem of the Berlin Academy was now capable of solution. My ambition at the time did not go further than this . . .[7]

Hertz had quickly realized that the phenomenon could only be due to the occurrence of oscillatory currents in the spiral coils but he further realized - and this was the truly significant step - that the period of the oscillations, 'estimated, it is true, only by the aid of theory', was 'of the order of a hundred-millionth of a second'. Oscillatory currents with frequencies of this order were about a hundred times more rapid than those which had previously been observed by Fedderson and others and it is not altogether surprising that they immediately gave rise in Hertz' mind to the possibility of resuming an attack on the problem that he had been compelled to set aside five years previously.

The earliest reference to his discovery of these very high frequencies occurs in his letter[8] to von Helmholtz dated 5 December 1886, a letter of particular interest in that it contains a simple sketch and description of what we would now call a 'Hertz dipole', and thereby indicates that in his experiments with these higher frequencies he has progressed from the coiled form of inductance - as represented by the Knochenhauer Spirals - to an open-wire linear form. Regrettably, he does not seem to have recorded the processes of thought which led him to make this exchange, an exchange for which, in the context of the times, the word 'invention' does not seem too strong a term. His letter, in part, reads:

> I have succeeded in demonstrating quite visibly the induction effect of one open rectilinear current on another rectilinear current, and I may hope that the way I have now found will enable me to solve one or other of the questions connected with this phenomenon.

> I produced the inducing rectilinear current in the following manner: A

Diagram 1 Inducing rectilinear current

> thick, straight copper wire, 3 m long, is attached at the ends to two spheres, 30 cm in diameter, or two conductors of similar capacity. There is a break in the midpoint of the wire for a spark gap of ¾ cm between two small brass spheres. Across them the crackling sparks of a large induction machine are allowed to pass, whereupon electric oscillations characteristic of the rectilinear circuit are excited (which, to be sure, was scarcely to be anticipated), and these oscillations now exercise a relatively strong inductive effect on the environs. I obtained sparks in a simple square circuit 75 cm on a side, consisting of thick copper wire and containing only a short spark gap, even at a distance of two metres from the induction path . . .

Almost 'in a stride', Hertz had not only discovered a radiating transmitter but a 'receiver' as well, very insensitive though it may have been!

His achievements at this stage were aptly summarized by Professor Hermann Ebert in the course of a Memorial Address[9] shortly after Hertz' death in 1894:

> (l) If we allow small condensers with small capacities to discharge through short and simply-constituted circuits, we obtain a sharply-defined discharge of very short duration, which is the long-sought-for sudden disturbance of electrical equilibrium. This was a discovery which could not be foreseen by any theory, a newly-discovered property of electric sparks, on which, however, all Hertz' later work depends. But theory showed further that at the same time success was assured on the second point, since oscillations of the

required period were obtainable; so that Hertz now had a powerful exciter of electrical vibrations.

(2) Such vibrations are capable of exciting in another discharge circuit of like form resonance phenomena of sufficient intensity, even when the two circuits are separated from each other by great distances.* Hertz had therefore found a receiver by means of which he could detect the waves produced by the exciter and propagated to any distance.

Armed now with a generator of oscillations having a convenient wavelength for experiments in a laboratory and with a means by which, with extreme care and patience, he could detect them, Hertz spent the winter of 1886-7 investigating the properties of electric oscillations.

Hertz' oscillator - he called it his 'primary conductor' - usually consisted of a stout copper wire having a span of the order of one to three metres, divided in the middle to include a spark-gap of about ¾ cm between two brass spheres 3 cms in diameter. The primary conductor or, as we would call it, the 'dipole', was energized by a large Rühmkorff induction coil with a mercury interrupter energized in its turn by six large Bunsen cells. In some experiments, Hertz would 'load' the ends of his 'dipole' with two zinc spheres whose position on the copper wire could be easily adjusted to provide a ready means of varying the frequency of the oscillations. It must be emphasized that throughout all his work Hertz' 'receivers' consisted of nothing more than squares or circles of copper wire, each square or circle incorporating an adjustable 'micrometer spark-gap'. By very carefully adjusting the length of the spark gap (within a tenth of a millimetre) to the point where the spark just failed to jump the gap, Hertz was able to assess the relative intensity of the currents induced into his various 'side circuits'.

While Hertz was conducting the long series of experiments with sparks, electrical oscillations and resonance during the winter of 1886-7, he made the unexpected observation that light had a marked influence on the length of spark which would occur in any given circumstances. If the spark gap in the primary discharge circuit of an induction coil was within sight of the micrometer gap in a nearby resonant circuit, Hertz observed that the spark obtainable across the micrometer gap was considerably longer than it was if he screened the micrometer gap in a darkened enclosure to facilitate the observations. The effect was so remarkable that it seemed to call for further investigation and resulted the following spring in the publication of a paper, 'On an effect of Ultra-Violet Light upon the Electric Discharge' in the leading journal, *Wiedemann's Annalen*.[10]

Hertz had no intention of allowing this phenomenon of ultra-violet light to divert his attention from his main line of research but, for some time, he was in

---

* Readers may care to be reminded that these words, 'great distances' and 'any distance' must be interpreted in the context of the times. In 1894 - i.e. pre-Marconi - and even more in 1887, a 'great distance' would have signified a distance of only a *very* few metres.

some doubt as to whether or not he was observing some new form of action at a distance. As soon as he knew for certain, however, that he was only dealing with an effect of ultra-violet light, he put aside its further investigation to resume the original course of his experiments. The publication of his paper in *Wiedemann's Annalen*, however, attracted widespread interest among contemporary scientists and ensured that any further paper coming from the young professor at Karlsruhe would be given close attention. In this somewhat indirect way it seems probable that Hertz's subsequent discoveries in the field of electromagnetic radiation received earlier and more widespread attention than they might otherwise have done.

Returning to the original course of his experiments, Hertz spent several months investigating what we would now describe as the electromagnetic field in the immediate vicinity of a dipole - and, when one remembers the extraordinarily primitive and insensitive nature of his 'detector', his results fill one with admiration. The patience he must have exercised is truly astonishing.

Quite apart from his very insensitive 'receiver', his 'transmitter' could not be relied upon to deliver an unvarying output. Anyone who has ever had occasion to make use of a Rühmkorff induction coil will be familiar with the vagaries of its operation and the irregularity of its output - usually due to the erratic behaviour of the make-and-break - so that the signal-level radiated by his dipole was liable to vary. Moreover, since his measurements of the pattern of the field-strength in the immediate vicinity of his dipole had to be based on his estimate - in tenths of a millimetre - of the length of a tiny spark in a 'micrometer spark-gap', the achievements recorded in his next paper, *On the Action of a Rectilinear Oscillation upon a Neighbouring Circuit,*[11] command highest admiration.

Of all Hertz' experimental researches, this was possibly the most difficult and certainly one of the most important, for it formed the basis of his discoveries during the next two years; discoveries which, in their ultimate effect, were to change the face of the civilized world. In reviewing his successive achievements, we need to bear in mind that *never,* throughout the whole course of his experiments, even until the very end of his life in 1894, was there any thought in his mind that this new range of electromagnetic frequencies which he had discovered would have even the smallest application as a means of communication.

If Hertz had no thoughts of radio communication, what exactly *were* his aims as he pursued the course of his experiments in his laboratory at Karlsruhe? To answer this question, we must remember that Hertz was a pure physicist with a passionate desire to work towards the increase of human knowledge - in his own words, 'to expand the frontiers of science'. He was an 'academic', pure and simple, with little or no interest in any commercial aspects. Early in his career he had been encouraged by von Helmholtz to devote his attention to electromagnetic problems and this had quickly led him to the need for a confirmation, or a refutation, of Maxwell's theory which differed so markedly from the current understanding of Continental scientists but still lacked any demonstrable proof after more than fifteen years. It had not taken young Hertz long to realize that if he was to demonstrate that - in Maxwell's words - 'light is an electromagnetic disturbance propagated through the field according to electromagnetic laws', he would need to be able to generate electromagnetic waves by means which were manifestly electromagnetic in nature, which were of a length convenient for experiment in a laboratory, and which could be demonstrated to possess all the properties of visible light, radiation, reflection, refraction, diffraction and, above all, identical velocity.

Happily, and as a direct outcome of his experiments with the Knochenhauer Spirals, he now possessed the means whereby he would be able to demonstrate that waves generated electromagnetically did in fact possess all the properties normally associated with waves of light. By doing so, he would confirm the validity of Maxwell's theory: *this* was the immediate object of his further experiments at Karlsruhe.

Hertz died on 1 January 1894, at the tragically early age of thirty-six, a few years too early to witness the practical development of his momentous discoveries. He left a young widow and two very young daughters. In 1936 Frau Elizabeth Hertz, by then an elderly lady who had been a widow for forty-two years, left her home in Bonn and, with her two daughters, Johanna, a Doctor of Medicine, and Mathilde, a Doctor of Physics, came to settle in Cambridge.

Figure 8   Frau Elizabeth Hertz (photograph by the author)

In 1938 the present author had the honour of receiving from the hands of Frau Hertz the original manuscript of the paper, *On Electric Radiation*.[12] Twenty years later in 1958 her daughter similarly handed over the manuscripts of nearly all Hertz' earlier papers, which now form part of the National Collection in the Science Museum. Frau Hertz died in 1941 and Dr Johanna in 1966, but Dr Mathilde survived until 1975, residing quietly in their house at Girton.

References

1. Included in H. Hertz, *Memoirs: Letters: Diaries*, arranged by Johanna Hertz, 2nd edition enlarged and edited by Mathilde Hertz and Professor Charles Susskind, San Francisco Press, 1977.
2. *ibid.*
3. H. Hertz, 'Experiments to determine an Upper Limit for the Kinetic Energy of an Electric Current', *Miscellaneous Papers*, Macmillan, London, 1896, **1**, No.4.
4. H. Hertz, 'On the distribution of electricity over the surface of moving conductors. 1881, *Miscellaneous Papers*, Macmillan, London, 1896, **1**, No.3.
5. H. Hertz, *Electric Waves* (translated into English by D. E. Jones), Macmillan, London, 1893; reprinted by Dover Publications Inc., New York, 1962.
6. H. Hertz, 'On the relations between Maxwell's Fundamental Electromagnetic Equations and the Fundamental Equations of the Opposing Electromagnetics', *Miscellaneous Papers,* Macmillan, London, 1896, **1**, pp.273-290.
7. Hertz, *Electric Waves*, p.2.
8. Hertz, *Memoirs*.
9. Professor Hermann Ebert, 'Heinrich Hertz: a Memorial Address delivered to the Physical Society of Erlangen on 7 March 1894', translated by J. L. Howard, *The Electrician*, 1894, **33**, pp.272-274, 299, 332-335.
10. H. Hertz, 'Ultra-violet Light upon the Electric Discharge', in *Wiedemann's Annalen*, 1887, **31**, p.983.
11. H. Hertz, 'On the Action of a Rectilinear Oscillation upon a Neighbouring Circuit', in *Wiedemann's Annalen*, 1888, **34**, p.155. See also Hertz, *Electric Waves*, pp.80-94.
12. Hertz, *Electric Waves*, pp.172-185.

Chapter 5

Lodge

Oliver Lodge (1851-1940)

The great contribution made by Sir Oliver Lodge (he was knighted in 1902) to the early development of wireless telegraphy has been very largely forgotten. In comparison with the seemingly more dramatic achievements of the youthful Marconi, Lodge's work held little of interest for the lay press around the turn of the century. It was well-known and appreciated by his contemporary peers, but they were perhaps too few in number to have ensured a lasting record in the public mind. Lodge's work was almost entirely scientific, Marconi's was solely devoted to development and to practical application: while Marconi's achievements received frequent and generous acknowledgement in the lay press, Lodge's contributions were virtually unnoticed. Yet without Lodge there would have been no Marconi. Not only did Lodge prepare the ground over at least a decade in which the seeds of Marconi's work were to germinate but, by providing independently a number of the essential elements, he made possible the development of Marconi's ideas into a practical system of wireless communication.

One of the objects of this book is to set in its correct perspective the contribution made by Lodge both before and after Marconi's appearance upon the scene. Without Lodge, the development of wireless telegraphy would have been long delayed and, with its delay, the whole history of the civilized world throughout the present century might well have been very different. Far in advance of any other, Lodge appreciated the need for resonance between transmitter and receiver.

Lodge's work, in fact, was not merely of academic and historic interest: it constituted a truly essential element in the early development of wireless telegraphy and it seems only proper that it should now be set in its true perspective.

Education

Oliver Lodge was born at Penkhall, near Stoke-on-Trent on the 12 June 1851, the eldest son of a small merchant who supplied the many pottery firms in the district with certain of their raw materials. After the custom of the time, he expected his sons to succeed him in the family business. Not for the first time, a father was to be disappointed in this ambition.

Writing of his school days in after years, Lodge himself has said that they were for him among the most miserable days of his life and although one may suspect that there was an element of exaggeration in this statement, the life of an average schoolboy in the middle of the nineteenth century was certainly not easy. His real education for a career in science unquestionably commenced at the age of fifteen with a visit to an aunt in London who encouraged him to attend a series of popular lectures on chemistry at the old College of Chemistry in Great Marlborough Street. Fascinated by the elementary but to him novel experiments, he needed little encouragement to attend further lectures on geology at King's College and on heat by Professor Tyndall of the Royal Institution at the Museum of Geology, then located in Jermyn Street. These lectures, especially those by Professor Tyndall, made a deep impression on the boy, who after several months in London returned home with dragging heels to assist his father in his business. It was already evident, at least to himself, that his real interests lay elsewhere.

Early in the 1860s the old 'Science and Art Department'* of South Kensington had instituted a series of educational courses in scientific subjects in a number of industrial towns throughout the midlands and one of these courses - on chemistry - was held at the Wedgwood Institute in Burslem, not far from the Lodge home. Enrolling for the course, young Lodge, who was then sixteen, soon found that the lecturer, a certain John Angel, had to travel from Manchester to give his lectures and having, in consequence, little time to prepare his experiments, was only too glad to accept the boy's services as an unpaid but enthusiastic assistant. In this manner and in his spare time from his father's business, Lodge learned a great deal of practical chemistry. In the following year, when the courses sponsored by the Science and Art Department were expanded to include others on heat, light, sound, electricity and mathematics, Lodge took full advantage of the opportunity to acquire a broad general knowledge of basic physics. In the ensuing examinations, Lodge took first place in each of the eight subjects in which papers were set, an achievement which led to some relaxation of the disfavour with which his father had hitherto regarded his scientific interests. Bowing perhaps to the inevitable, his parents permitted

---

* The Department of Science and Art, established in 1853, was a government department operating under a committee of the Privy Council known as the Committee of Council of Education which had been established in 1839. One of its principal objects was the promotion of secondary education and the training of teachers throughout the country. In 1899 the Science and Art Department was merged with the Education Department to form the Board of Education.

OLIVER JOSEPH LODGE

Figure 9   Oliver Joseph Lodge

him to enrol during the following winter for a series of courses by the Science and Art Department at South Kensington, primarily intended for the better education of teachers throughout the country.

Lodge worked hard at South Kensington during the winter of 1869-70, studying biology, physics and chemistry under Professors Huxley and Frankland. Disappointed at finding no courses in mathematics at South Kensington, he enrolled for special evening courses in mathematics and physics at King's College under W. G. Adams. It is evident from his autobiography, *Past Years*,[1] that Lodge was a self-motivated and very hard-working student. For two years he contrived to sandwich his studies with intermittent participation in his father's business. His remarkable achievement in being the only student to win first class honours in physics in London University's Intermediate B.Sc. Examination in 1873 finally persuaded his father to relent and allow his son to pursue the course upon which he was so clearly determined.

Enrolling at University College London for the 1873-4 session, Lodge studied pure mathematics under Professor Henrichi, applied mathematics under Professor Clifford and physics under Professor Carey Foster. Lodge's deep interest in Maxwell's theory was probably first stimulated by his good fortune in hearing

Maxwell speak at the Bradford meeting of the British Association in 1873. At the time Maxwell was forty-two, the Cavendish Professor of Physics at Cambridge and already a Fellow of the Royal Society, while Lodge, twenty years younger, was still a student. By the time Lodge received his B.Sc. degree in 1875, he was fully conversant with Maxwell's theory and the mathematical basis upon which it was founded, but we must remember that in 1875 Maxwell's theory was a theory only, warmly received by many of the mathematical physicists of the day, yet lacking that demonstrable proof which alone could bring it universal acceptance.

While continuing to study for the D.Sc. degree upon which he had set his sights and which he was to receive in 1877, Lodge was offered a post as assistant and demonstrator under Professor Carey Foster, thereby becoming a member of the staff of the Physics Department. Additionally, he accepted a part-time post teaching physics at Bedford College for Ladies, a post which he continued to hold until 1881 when he was appointed Professor of Experimental Physics at the newly formed University College of Liverpool.

While waiting to take up his Liverpool appointment, Lodge undertook an extensive European tour primarily to make personal contact with leading continental scientists to whom he carried letters of introduction. Calling without notice on the great Professor Helmholtz in Berlin, Lodge found him on the point of going out. After a few words, Helmholtz handed him over to his demonstrator, Dr Heinrich Hertz, then a young man of twenty-four, with whom Lodge quickly established a friendship which was to be closely maintained until Hertz's untimely death in 1894.

At the time of Lodge's appointment, the only building available to the new University College at Liverpool was an old lunatic asylum which required a great deal of alteration and adaptation before it could adequately fulfil its new function. Fully occupied by his task of building up a Department of Physics in a new university, some years were to elapse before Lodge could again give serious thought to the problems presented by Maxwell's theory, although it is clear that the subject was never far from his mind.

For about twenty years Maxwell's theory was a theory only and two great problems existed. The first was to find a means for generating the waves; the second, far harder, was to find the means whereby they could be detected and their physical existence demonstrated. It had been known for many years that, under suitable conditions, the discharge of a Leyden jar was oscillatory - Henry had shown as much in 1842, as we saw earlier - but it was Professor G. F. FitzGerald of Dublin who in 1883 first suggested that such oscillations ought to emit electromagnetic waves of radiation - in a manner somewhat similar to that in which the vibrating arms of a tuning fork emit sound waves in air. Oliver Lodge, who was on terms of close personal friendship with FitzGerald, agreed completely on mathematical grounds that FitzGerald's suggestion was sound but, as he was to say later, 'it is in fact ridiculously easy to produce the waves; the difficulty was to find evidence of them'. In the absence of any means of detection, further progress seemed halted.

Lodge and lightning

At this stage it is necessary to digress somewhat to take account of Lodge's involvement with the subject of lightning, an involvement which was to have far-reaching consequences in the years to come.

In earlier times, the real nature of lightning was a mystery, its destructive force the origin of much fear and superstition. The hazardous experiments of Benjamin Franklin in 1751 with his high-flying kite may have done little to *explain* the phenomenon but at least they proved beyond doubt that a lightning flash was nothing other than a very large electric spark. This had provided the basis for a somewhat pragmatic development of the practice of installing 'lightning conductors' on high buildings. Unfortunately they did not invariably provide the degree of protection desired and severe damage continued occasionally to arise, particularly in those countries where violent thunderstorms were prevalent.

It was evident that the whole subject required thorough investigation by a competent scientist and towards the end of 1887 Lodge was invited to deliver two lectures on the subject to the members of the Society for the Encouragement of Arts, Manufactures and Commerce (usually known as the Society of Arts). In preparation for these lectures, Lodge undertook a long series of experiments on the discharge of Leyden jars which, as Henry had shown in 1842, were often of an oscillatory nature. Lodge's experiments showed beyond doubt that a lightning flash was oscillatory and, in consequence, it was evident to him that the protection of a building from the effects of a strike was by no means a simple matter. It was not sufficient merely to provide a path of low resistance to carry the discharge to earth - as was assumed by the 'electricians' of the day: the self-inductance of the conductor was of far more importance than its resistance and, unless the self-inductance was low, the lightning would often prefer to jump through the air rather than follow the apparently easy path through the conductor. In the course of the experiments with the Leyden jars, Lodge not only demonstrated the existence of electromagnetic waves but, as we shall see, he also showed the need for resonance between oscillatory circuits if an adequate transfer of energy was to be achieved between them.

Lodge's first lecture to the Society of Arts was entitled *The Protection of Buildings from Lightning.*[2] It was delivered on 10 March 1888 and was mainly confined to a consideration of the phenomenon of lightning itself. He drew attention to the similarities between a lightning flash and the discharge of a Leyden jar, he criticized some of the then current views on the construction of lightning conductors, and he concluded by drawing emphatic attention to the need for lightning conductors to have a minimum of self-inductance, this factor being of far more importance than the resistance.

In his second lecture, delivered a week later, Lodge demonstrated a considerable number of experiments with Leyden jars. He referred to the analogy between mechanical inertia and self-induction in electrical circuits and demonstrated that the circumstances of a discharge are regulated 'far more by

inertia than by conductivity'.   In one experiment, he showed a Leyden jar
discharging across a spark gap nearly 1½ inches in length, although the gap was
short-circuited by a loop of stout copper wire 40 inch in length which had a
resistance of only 0.025 ohm.   Such experiments emphasized the enormous
importance of inductance or 'electrical inertia' where oscillatory currents were
concerned.   In passing, Lodge referred to the fact that rapidly oscillatory currents
travelled only on the surface of a conductor.   'In the outer skin, of microscopic
thickness, electricity will be oscillating to and fro, but the interior of the
conductor will remain solidly inert and take no part in the action'.

In another of the many experiments which Lodge performed at the second
lecture, he described an effect to which he gave the name 'recoil kick'. He had
arranged a pair of Leyden jars to discharge through a gap A (Diagram 2) but,
with a pair of long wires attached to each side of the gap, he found that every
time the jars were discharged at A, an even longer spark would take place
between the gaps along the pair of wires, the longest spark being obtained at the
far end of the wires. Lodge observed:

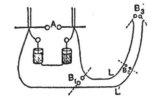

Diagram 2 Recoil kick

The question arises why this be so. Plainly what is happening is this: the
discharge at A sets up electrical oscillations . . . the electricity in the long
wires is surging to and fro like water in a bath when it has been tilted and the
long spark at the far end of the wires is due to the recoil impulse or kick at
the reflection of the wave . . . The nearer the length of the conductors
corresponds to a half-wavelength, or some multiple of half-wavelength, of the
oscillations produced by the discharging jars, the more perfect will be the
synchronism between the pulses, and a longer recoil kick may be expected.[3]

In other experiments shown at the same lecture, Lodge not only demonstrated the
physical existence of electric waves along wires but he was able to make close
estimates of their actual length. In a footnote to the printed report, he remarked:

Since the delivery of the lecture, a great number of quantitative observations
on these lines have been made. Evidence of electro-magnetic waves 30 yards
long has been obtained. I expect to get them still shorter.[4]

Although he could demonstrate the existence of electric waves along wires and measure their wavelength, he had no means of detecting waves in space. In layman's terms, he was in much the same position as anyone today who, while knowing that the air all around is full of radio signals, is totally unable to detect their presence without some form of receiver. Lodge, as yet, had no receiver.

Later in the same year - 1888 - Lodge published a more mathematical paper, 'On the Theory of Lightning Conductors', in the *Philosophical Magazine*.[5] In the opening paragraph he acknowledged the debt to Sir William Thomson (later Lord Kelvin) for his important paper in 1853,* 'On Transient Electric Currents',[6] which showed how to calculate the wavelength of an electrical oscillation in terms of the inductance and capacity of the circuit. Lodge further acknowledged the debt to Oliver Heaviside for pointing out that rapidly alternating currents confine themselves to the exterior of a conductor. He then proceeded to apply these considerations to the case of a lightning discharge.

Towards the end of this paper, Lodge referred again to the 'Experiment of the Recoil Kick', which he had first shown to the Society of Arts some months earlier. In explanation of the Diagram 3, he wrote:

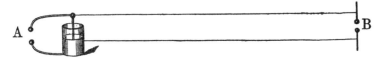

Diagram 3 A discharging condensor circuit.

The jar discharges at A in the ordinary way and simultaneously a longer spark is observed to pass at B at the far end of the two long leads . . . The theory of the effect seems to be that oscillations occur in the A circuit according to the equation:

$$T = 2\pi \sqrt{(LS)}$$

where L is the inductance of the A circuit and S is the capacity of the jar. These oscillations disturb the surrounding medium and send out radiations, of the precise nature of light, whose wavelength is obtained by multiplying the above period by the velocity of propagation . . . [7]

He continued to discuss in mathematical terms, the conditions for the propagation of these oscillations along the wires towards B and deduced that: 'the velocity of propagation of condenser discharges along two parallel wires is simply the velocity of light, the same as in general space'. The paragraphs which follow are of importance in that they show in the clearest possible way that at this date - mid-1888 - Lodge was conversant with all the essential principles of tuning and electrical resonance:

* The date was incorrectly given by Lodge as '1858'.

The pulses rush along the surface of the wires, with a certain amount of dissipation, and are reflected at the distant ends, producing the observed recoil kick at B. They continue to oscillate to and fro until damped out of existence exponentially due to the power lost in resistance and radiation. The best effect should be observed when each wire is half a wavelength, or some multiple of half a wavelength, long. The natural period of oscillations in the wires will then agree with the oscillation period of the discharging circuit, and the two will vibrate in unison, like a string or a column of air resounding to a reed.

Hence we have here a means of determining experimentally the wavelength of a given discharging circuit. Either vary the size of the A circuit, or adjust the length of the B wires, until the recoil spark at B is as long as possible.

The paper in the *Philosophical Magazine* was dated 'University College, Liverpool, 7th July 1888', but before it was set in print, Lodge forwarded a postscript from Cortina in the Tyrol where he had gone on holiday:

Since writing the above, I have seen in the current July number of *Wiedemann's Annalen* an article by Dr Hertz, wherein he establishes the existence and measures the length of aether waves excited by coil discharges; converting them into stationary waves, not by reflexion of pulses transmitted along a wire and reflected at its free end, as I have done, but by reflexion of waves in free space at the surface of a conducting wall . . . The whole subject of electrical radiation seems to be working itself out splendidly.

Cortina, Tyrol, July 24, 1888

The news of Hertz' discoveries was warmly welcomed by Lodge and by his friend, Professor G. F. FitzGerald, the two men who, by virtue of their long interest and deep understanding of the whole subject, were probably better qualified than any others to appreciate the full significance. The meeting of the British Association in Bath during September 1888 provided an opportunity to call attention to Hertz' researches and, as has already been mentioned, FitzGerald, in his Presidential Address[8] to Section A, gave a broad and generous description of them. Lodge, who at the same meeting read two papers of his own, 'On the Measurement of the Length of Electromagnetic Waves' and 'On the Impedance of Conductors to Leyden jar Discharges', paid high tribute to the work of Hertz and his outstanding achievements but, like Hertz himself, clearly saw no practical use for the discoveries. Both at the time regarded them merely as providing a superbly interesting confirmation of what had hitherto existed only in the realms of abstruse mathematical theory. Although the reports of Hertz' discoveries attracted wide interest among physicists, it is perhaps not surprising that in the absence of any practical application, they aroused little if any interest in the mind of the general public.

Lodge and Preece

Lodge's theory that a lightning flash consisted of an oscillatory discharge and was not just a simple discharge of current in one direction, as had previously been assumed, was not received with the universal acclaim to which it was certainly entitled. Indeed, it would be no exaggeration to say that it gave rise to a bitter controversy in which the principal participants were Lodge himself and the 'electricians', led by William Preece of the Post Office. The electricians prided themselves with being 'practical men' of vast experience in the design and erection of lightning conductors over a period of many years and much of their criticism was based upon the fact that Lodge had developed his theory around a few simple laboratory experiments in which he employed sparks of only an inch or two in length. They argued that he was quite unjustified in extrapolating the results of such experiments to embrace a discharge with the dimensions of a lightning flash.

The controversy, which raged for several weeks during the summer of 1888 in the columns of the technical press and in *The Times*, was one with all the elements of a dispute between major contestants with irreconcilable views. On the one side Lodge, with all the evidence of his scientific experiments in support of his theory, had no practical experience with lightning conductors; on the other Preece, the chief electrical engineer of the Post Office, nearly twenty years older than Lodge, claimed near infallibility on the grounds of having nearly half a million lightning conductors in his charge.

The managers of the British Association, anticipating a lively debate, decided to hold a full-scale discussion on lightning conductors at the annual meeting in Bath during early September 1888. The main participants were, of course, Preece and Lodge, and their total disagreement quickly became evident to the large audience. The heights of rhetoric attained during the three-hour discussion may not have reached those of the famous debate between Huxley and Bishop Wilberforce at an earlier meeting, but the degree of bitterness between the main participants and the total failure on the part of Preece even seriously to consider Lodge's theory were clear indications that any future understanding between them was unlikely.

The importance of this dispute between Lodge and Preece lies not in its contemporary connection with the subject of lightning conductors but with the explanation it provides of Preece's treatment of Marconi when, eight years later, the young man arrived in England at the age of only twenty-two. As we shall see later, the generously paternal manner in which Preece, then a man of sixty-two, received young Marconi and the way in which, at least initially, he placed the Post Office facilities at his disposal has always been a matter of some curiosity, particularly on account of the influence it exerted on the way in which Marconi proceeded with the early development of his ideas for a system of wireless telegraphy. But with relations between Preece and Lodge as strained as they clearly were, Preece's attitude towards Marconi becomes easier to understand. As we shall see later, it seems not unlikely that this clash between Preece and

Lodge was largely responsible for Marconi's progress taking a somewhat pragmatic course instead of undergoing a period of development at the hands of some suitably qualified scientist as one might have expected.

Lodge and electromagnetic oscillation

The influence of Hertz' researches on Lodge's own work is plainly evident in a Discourse which Lodge delivered to the members of the Royal Institution at a Friday evening meeting on 8 March 1889. In this lecture on 'The Discharge of a Leyden Jar',[9] Lodge repeated and amplified a number of the experiments he had shown to the Society of Arts the previous year, but he especially emphasized this time the oscillatory nature of the discharge, the manner in which the oscillations decay and the phenomenon of 'sympathetic resonance'. In particular, he dwelt upon the phenomenon of radiation.

By long tradition, the audience at a Friday Evening Discourse at the Royal Institution is not expected to be highly specialist. Lodge therefore presented his subject in non-mathematical terms and the text, as printed in the *Proceedings*, constitutes a superb survey of the 'state of the art' as it existed in the spring of 1889. After describing analogies between mechanical vibrations and the oscillations in an electrical circuit, Lodge referred to the conditions which determine the rate at which oscillations die out. Resistance, corresponding to mechanical friction, was important but:

> there is another cause, and that a most exciting one. The vibrations of a reed are damped, partly indeed by friction and imperfect elasticity but partly by the energy transferred to the surrounding medium and consumed by the production of sound. It is the formation and propagation of sound which largely damps the vibrations of any musical instrument. So it is with electricity. The oscillatory discharge of a Leyden jar disturbs the medium surrounding it, carves it into waves which travel away from it into space; travel with a velocity of 185,000 miles a second; travel precisely with the velocity of light.
>
> The second cause, then, which damps out the oscillations in a discharge circuit is *radiation* which differs in no respect from ordinary radiation or 'radiant heat' and it differs in no respect from light except in the physiological fact that our eyes respond only to waves of a particular, and that a very small, size while radiation in general may have waves which range in length from 10,000 miles to a millionth of an inch.

Passing on to describe the discovery of the oscillatory nature of the discharge of a Leyden jar by Joseph Henry in 1842, Lodge continued:

> One immediate consequence and easy proof of the oscillatory character of a Leyden jar discharge is the occurrence of [the] phenomena of sympathetic resonance.
>
> Everyone knows that one tuning fork can excite another at a reasonable distance if both are tuned to the same note. Everyone knows, also, that a

fork can throw a stretched string attached to it into sympathetic vibration if the two are tuned to unison or to some simple harmonic. Both these facts have their electrical analogue . . .

A Leyden jar discharge can so excite a similarly-timed neighbouring Leyden jar circuit as to cause the latter to burst its dielectric if thin and weak enough. The well-timed impulses accumulate in the neighbouring circuit till they break through a quite perceptible thickness of air.

Put the circuits out of unison by varying the capacity or by including a longer wire in one of them; then, although the added wire be a coil of several turns, well adapted to assist mutual induction as ordinarily understood, the effect will no longer occur until the capacity is suitably diminished and the synchronism thus restored.

That is one case, and it is the electrical analogue of one tuning fork exciting another . . .

The other case, analogous to the excitation of a stretched string of proper length by a tuning fork, I published last year under the name of the experiment of the recoil kick, where a Leyden jar circuit sends waves along a wire connected by one end with it, which waves splash off at the far end with an electric brush or long spark.

The above extracts from his discourse to the Royal Institution provide ample evidence that Lodge in the early spring of 1889 was well acquainted with the basic principles of electromagnetic radiation and, in particular, with the vital part played by resonance in the transfer of energy between one electrical circuit and another.

The whole subject was more fully developed in a paper, 'On Electrical Radiation and its concentration by Lenses',[10] which Lodge read to the Physical Society on 11 May 1889, subsequently published in the *Philosophical Magazine* in July of the same year. In the paper, Lodge described the experiments which he and his assistant, Dr Howard, had carried out in extension of Hertz' work during the previous year. Hertz had shown that electromagnetic waves behaved like rays of visible light and could similarly be reflected and refracted; Lodge argued that it must similarly be possible to concentrate and focus them by means of suitable lenses and he proceeded to demonstrate this property by the use of a large pair of plano-convex cylindrical lenses of common mineral pitch in conjunction with a Hertzian oscillator which emitted electromagnetic waves of about one metre wavelength.

In the course of the paper, Lodge digressed to make a number of points of considerable significance in the light of developments in the practice of actual wireless telegraphy some eight or nine years later. Referring to the form of an oscillatory circuit, he showed that a half-wave linear oscillator - the now familiar 'Hertz dipole' - is a far more efficient radiator than the large closed circuits which he had been using in his Leyden jar experiments. 'For obtaining distant effects', Lodge wrote, 'the linear oscillator is vastly superior. Its emission of plane-polarized, instead of circularly polarized radiation is also convenient . . .'.

While the Hertzian oscillator is so much better as a *radiator,* Lodge pointed out that *because* it is so efficient as a radiator, the oscillations die away extremely quickly: 'the damping out of the vibrations is so vigorous that all oscillations after the first one or two are comparatively insignificant.' With closed circuits, however, Lodge pointed out that the oscillations are sustained very much longer. Because the oscillations are so little damped by radiation, they tend to persist with but little reduction in amplitude and, as in a mechanical system where oscillations are long sustained, the phenomenon of resonance between two circuits tuned to the same frequency is far more conspicuous. In brief, Lodge declared that you can have an open circuit, such as a Hertz dipole, which is a good radiator but a poor resonator, or you can have a closed circuit which is inefficient as a radiator though it may be an excellent resonator; what you *cannot* have is a circuit which is simultaneously a good radiator and a good resonator. It is of course necessary to remember that Lodge was writing at a time when the only means of generating high frequency oscillations was by the use of the spark discharge of a Leyden jar or an induction coil.

Finally, in an interesting passage in his paper to the Physical Society in May 1889, Lodge deals with the power radiated by the small Hertzian dipole which he employed for his lens experiments. Assuming that his induction coil charged the two halves of the oscillator to a potential difference of the order of 26,400 volts, he calculated that the power of the initial radiation was almost 100 kW. But, as he pointed out, almost the whole stock of energy is dissipated in the first one and a half swings at each discharge and:

> Nothing approaching continuous radiation can be maintained at this enormous intensity without the expenditure of great power - in this case an hundred and thirty horse-power if my calculation is correct . . . To try to make the radiation more continuous a large induction coil excited by an alternating machine of very high frequency might be tried. But even if sparks were made to succeed each other at the rate of 1,000 per second, the effect of each would have died out long before the next one came. It would be something like plucking a wooden spring which, after making 3 or 4 vibrations, should come to rest in about two seconds and of repeating the operation of plucking regularly once every two days.

The several lectures and publications by Lodge which have been referred to above were all addressed to the comparatively small circles represented by scientific societies, but with the publication early in 1890 of a letter in *Nature*[11] in which he described a simple experiment to illustrate electrical resonance, he introduced the whole subject to a much wider audience. The experiment he described was in fact very similar to one he had shown at the Royal Institution nearly a year earlier but in consequence of its publication in *Nature* it became far more widely known and in view of the influence it exerted, it is perhaps permissible to quote it in full:

Easy Lecture Experiment in Electric Resonance

An experiment, exhibited by me in its early stages at the Royal Institution a year ago, and since shown here in various forms, on the overflow of one Leyden jar by the impulses accumulated from a similar jar discharging in its neighbourhood, is so simple an illustration of electric resonance, and so easily repeated by anyone, that I write to describe it.

Two similar Leyden jars are joined up to similar fairly large loops of wire, one of the circuits having a spark-gap with knobs included, the other being completely metallic, but of adjustable length. The jar of this latter circuit has also a strip of tinfoil pasted over its lip so as to provide an overflow path complete with the exception of an air-chink, C. It is important that this overflow path be practically devoid of self-inductance. A jar already perforated could well be utilized for the purpose.

Then if the two circuits face each other at a reasonable distance, and if the slider, S, is properly adjusted, every discharge of A causes B to overflow. A slight shift of the slider puts them out of tune.

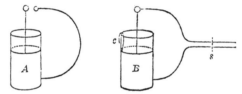

Diagram 4 Lodge's easy lecture experiment in electric resonance

Instead of thus adjusting by variable self-inductance, my assistant, Mr Robinson, has made a slight modification by using a condenser of variable capacity, consisting of two glass tubes coated with tinfoil, one sliding into the other, and joined by a flexible loop of wire; an easy overflow from one coat to the other being likewise provided. On making this loop face the discharging circuit of an ordinary Voss machine with customary small jars *in situ*, bright sparks at the overflow gap occur whenever the common machine sparks are taken, provided the sliding condenser be adjusted to the right capacity by trial.

There is little or no advantage in using long primary sparks: the vibrations are steadier and more definite with short ones. It is needless to point out that the two jars constitute respectively a Hertz oscillator and receiver, but fair precision of timing is more needed with these large capacities than with mere spheres or discs, because the radiation lasts longer and there are more impulses to accumulate. Hence actual resonance as distinguished from the effect of a violent solitary wave is better marked. Moreover, the sparks are bright enough to be easily seen by a large audience.

University College
Liverpool
20th February 1890

Oliver J. Lodge

For many years the experiment described in this letter continued to be a favourite in lecture theatres and thousands of students must have received their first vivid impressions of electrical resonance by seeing it performed. It was by no means an obvious experiment since it was, at first sight, difficult to understand how a Leyden jar, apparently short-circuited by a stout copper wire, could possibly become sufficiently charged to 'overflow', but if the circuit is re-drawn in a modern form, any radio engineer with experience of the ways of continuous-wave transmitters will immediately recognize the tendency of a capacitance in a nearby

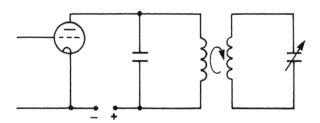

Diagram 5   Lodge's 'lecture experiment' re-drawn in modern form

circuit to flash-over as the circuit is tuned to resonance with the transmitter. In the days when Lodge first performed the experiment however, there were few indeed who understood the cumulative effect of successive oscillations or who recognized that even the most minute break in the circuit would prevent the building up of oscillatory currents and the development of a high-frequency voltage of sufficient magnitude to produce a flash-over. In fact this simple experiment probably did more than anything else to fix in the minds of those studying this branch of physics the elements of resonance phenomena in electrical circuits.

Lodge and the coherer

Although Lodge had narrowly missed the discovery of electromagnetic waves in space, his profound understanding of the essential principles of resonance in electrical circuits was to be of great significance. Of even greater importance, however, especially in the very early phases of practical radio communication was his work on the development of the 'coherer', that extraordinarily sensitive device for detecting electromagnetic waves generated by a spark discharge.

The phenomenon of cohesion between metallic surfaces had first come to his notice in 1889 when working on the protection of telegraphic instruments and cables from lightning. In the course of this work he observed that, when two metallic surfaces were separated by a minutely small air gap, they frequently became fused together when an electric discharge occurred in the neighbourhood. During 1891-2 he employed this principle in several different forms and, as he said, 'this arrangement, which I call a "coherer", is the most astonishingly sensitive detector of Hertzian waves'.[12]

During 1893 his attention was called by Professor G. M. Minchin[13] to the work of Edouard Branly in Paris. Branly had found that metallic dust or filings ordinarily possessed a very high resistance to the passage of an electric current but that, if even a small electric spark occurred in the vicinity, the resistance would fall to a low value. It is doubtful whether Branly recognized at the time that the effect was due to electromagnetic radiation but his observations were thorough and conclusive. He measured the reduction in the resistance of a whole range of substances in the form of powder or filings and noted that the effect was still marked even if the spark occurred at a distance of several yards.[14]

Lodge at once proceeded to try a tube of metallic filings as described by Branly and found it immeasurably superior as a detector to the cohering knobs and delicately balanced spring contacts he had been using hitherto. During the next few months he made many experiments with different forms of the filings coherer and with the aid of this new detector he carried out a long series of quasi-optical experiments which were shown to students and to members of the Liverpool Physical Society. Many of the coherers which were made at this period by Lodge and his assistant, E. E. Robinson, have survived and are now preserved in the collections of the Science Museum in London. One of them - filled with the original iron turnings - was used during the course of a television programme during 1960 and was found still to be amazingly sensitive, the resistance falling from many thousands of ohms to only a few hundred when an electric spark occurred nearby.[15]

A few months after Hertz' death on 1 January 1894, Lodge was invited to deliver a memorial lecture on 'The Work of Hertz' at the Royal Institution. This lecture can be seen in retrospect to have had tremendous influence[16] in disseminating an understanding of the properties of Hertzian waves beyond the small circle of mathematical physicists to whom the subject had appealed hitherto. The title of the lecture was a misnomer: in fact Lodge dealt more with his own work, and that of others who had followed Hertz, than with that of Hertz himself.

Two factors were probably responsible for the very considerable influence of this lecture on future developments. Firstly, Lodge described in detail the steps by which he had developed the coherer and, while remarking that 'this arrangement, which I call a "coherer", is the most astonishingly sensitive detector of Hertzian waves', he provided, in effect, very simple and precise instructions whereby such detectors could readily be duplicated, even by unskilled hands. Secondly, the text of his lecture was given very wide circulation by being reprinted in serial form through several consecutive issues of the weekly journal, *The Electrician*. It was further reprinted with additional material in the form of a small book entitled *The Work of Hertz and some of his Successors*.[17]

The lecture stimulated one who listened to it, the eminent telegraph engineer, Dr Alexander Muirhead, to propose that Hertzian waves might usefully be employed for the purposes of communication. This suggestion led directly to the close collaboration between Lodge and Muirhead which was to last for almost twenty years.

A few months later Lodge was requested to repeat his lecture at a meeting of the British Association for the Advancement of Science at Oxford in August 1894. In the meantime, Muirhead had provided him with a range of telegraphic equipment which included a Kelvin marine galvanometer and a Morse inker in order to show the audience the transmission of letters of the alphabet by means of electromagnetic waves. Before the lecture, Lodge installed a Hertz oscillator and an induction coil having a key in the primary circuit in the Clarendon Laboratory, a distance of about sixty yards from the lecture theatre. On the demonstration table in front of the audience, the receiving apparatus consisted simply of his coherer - a glass tube of metal filings - in series with a single cell and the marine galvanometer. When the key in the primary circuit was pressed for a short time by an assistant in the Clarendon Laboratory, a small deflection of the galvanometer took place in the lecture theatre, but when the key was pressed for a longer period, a larger deflection could be observed. Lodge was thus able to transmit dot and dash signals and so to demonstrate for the first time the possibility of sending messages by means of Hertzian waves.[18]

To those who heard the lecture in Oxford and witnessed the transmission of signals through several brick walls and across an open space, the achievement must have seemed miraculous, and it is not surprising that it was widely reported. What is perhaps surprising is that *at that time* no one seems to have considered what might have been the maximum range at which the signals could be detected, or whether there might be any practical application for a system of 'telegraphy without wires'.

With such profound knowledge of the fundamental theory of electromagnetic waves and with so much original experience and experimentation in their generation and detection, it has sometimes seemed surprising that Lodge did not himself proceed with their commercial development. One must remember, however, that he had heavy duties as the Professor of Experimental Physics at University College, Liverpool, and he could scarcely be expected to give up this career for the sake of developing a laboratory experiment and its commercial application, however promising it might seem. In later years he was to admit that he did not realize the commercial possibilities at the time but that, even if he had done, his decision would have been the same. 'It was stale news to me and to a few others', he wrote many years later, of the early demonstrations given by Marconi in 1896, 'but whereas we had been satisfied that it *could* be done, Marconi went on enthusiastically and persistently till he made it a practical success.'[19]

It may be appropriate at this stage to interpose a brief reference to the relations between Lodge and Marconi during the years following Marconi's arrival in this country early in 1896 and to speculate upon the very different course which early wireless telegraphy might have followed if the two men had agreed to collaborate from the outset instead of virtually ignoring the existence of each other. To appreciate the relationship properly, the very different circumstances of the two men should be taken into account for, although the actual course of events was unfortunate, it was perhaps not entirely inevitable.

By 1896 Lodge was a man of forty-five, a brilliant mathematical physicist who had been the Professor of Experimental Physics at Liverpool University for almost fifteen years. It is probable that in the 1890s he possessed a more thorough understanding of electromagnetic wave radiation than any other man alive.

In contrast, Guglielmo Marconi was a young man of only twenty-two, an Italian citizen but no stranger to England. He both spoke and wrote perfect English. He had spent nearly eighteen months experimenting with Hertzian waves on his father's estate, where he claimed to have sent signals over distances of up to a mile and a half. But having only a superficial understanding of the underlying physics, he was far more interested in achieving practical results than in the theoretical background which might lie behind them. In effect, he was what we would now term a youthful entrepreneur, lacking an academic or scientific background but brimful with energy, enthusiasm and patience, and gifted with a personality which enabled him to penetrate with comparative ease the upper echelons of official departments.

The course of events was to lead him to William Preece, the Chief Engineer of the General Post Office, a man who unfortunately was far more ignorant than he should have been of the developing science of electromagnetic wave radiation, of the work of Hertz and, in particular, that of Oliver Lodge. There seems little excuse for Preece's ignorance: Lodge's lectures on Hertz' discoveries and on his own experiments had been very widely reported during the previous year, both in the lay and technical press. Had he been aware of this previous work, it seems very unlikely that Preece would have sponsored young Marconi to the extent that he did. It was, in fact, almost certainly Preece's public 'adoption' and adulation for the young Marconi which caused Lodge, who held little regard for Preece, to regard the young man's activities with a measure of disdain.

By his failure initially to follow up the practical possibilities of his own researches, Lodge lost the opportunity to acquire popular acclaim as the 'inventor' of wireless telegraphy but his influence on its early development was nevertheless profound. With his deep understanding of the phenomenon of electrical resonance, he was the first to emphasize the importance of tuning - or 'syntony', as it was then called - and his patent No.11,575, for which he applied in May 1897 (and which provided for selective tuning by means of added inductance), was to become one of the most famous and fundamental patents in the whole history of radio communication. In passing, it is of interest to note that Lodge applied for this patent almost two months *before* the publication of Marconi's original patent.

It would be wrong to imply - as the above paragraphs may have done - that Lodge took no part whatever in the commercial development of wireless telegraphy. His association with Dr Alexander Muirhead in the Lodge-Muirhead Syndicate provided competition of some consequence for the Marconi Company during the early years of the twentieth century. Lodge himself was heavily pre-occupied with his academic duties at Liverpool and, from 1900 onwards, at Birmingham where he had been appointed Principal of the new University, yet he found time to co-operate with Muirhead in experimental work and by 1903 the Lodge-Muirhead system of wireless telegraphy had reached a degree of perfection which enabled General Greely of the US Signal Corps to write that the Lodge-Muirhead equipment:

> has proved more satisfactory than any other tested by the Signal Corps. Its primary point of superiority lies in the extreme regularity with which it functions . . . and in the beauty of workmanship and mechanical solidity . . .[20]

A full description of the Lodge-Muirhead System appeared in *The Electrician*[21] in March 1903 and in this account the influence of Muirhead's experience as a telegraph engineer is clearly apparent. The equipment was manufactured with the traditionally high standard of workmanship which had long been associated with Muirhead telegraph and cable apparatus and it is evident from the description in *The Electrician* that accuracy and reliability had been among the primary aims of the designers.

An important feature of the Lodge-Muirhead system, and one which undoubtedly contributed to the reliability of the receiver, was the rotating disc coherer invented by Lodge in 1902. It consisted of a small knife-edged steel disc which rotated slowly while just dipping into a globule of mercury upon which floated a thin film of oil. It was probably more sensitive, and certainly far more stable, than the filings coherers which had been employed hitherto.

An interesting commentary on the working of the Lodge-Muirhead system between the Andaman Islands and Burmah, a distance of 395 miles, appeared in *The Electrician*[22] in April 1906. In this account, it is mentioned that during the five months from July to November 1905, a total of 115,572 words was transmitted at speeds of between seventeen and twenty words per minute. 'The present installation', it was stated, 'must be regarded as fully reliable as an ordinary land-line of about 300 miles through forest . . . '.

It is thus evident that the Lodge-Muirhead equipment formed a thoroughly well engineered and practical system. Unhappily, the adoption of the system in this country was frustrated by the virtual monopoly which had been created by an agreement between Lloyds and the Marconi Company in 1901, whereby only Marconi apparatus would be used for a term of fourteen years at the many Lloyds signalling stations which then ringed our coasts. This agreement, together with the refusal of the Postmaster General to issue licences to competing stations, severely restricted the commercial field for systems other than Marconi's and, although the Lodge-Muirhead system found a limited application in the military field and overseas, the potential market was small.

It may have been partly this lack of a ready 'home market' and partly that both Lodge and Muirhead were well into their fifties and heavily pre-occupied with other commitments, but the fact remains that the Lodge-Muirhead system of wireless telegraphy failed to achieve the commercial success which its technical excellence would clearly have justified.

The final chapter of Lodge's pioneering work towards the development of radio-communication was not written until 1911 when an application for an extension of his master patent of 1897 came before Mr Justice Parker.[23] The validity of the patent was upheld, Lodge's claim to priority in the principles of resonance and selective tuning was established, and the exceptional extension of seven years was granted, mainly on the grounds that the inventor had been inadequately remunerated in consequence of the virtual monopoly created by the Marconi-Lloyds agreement and by the refusal of the Postmaster General to grant a licence for the working of the system in this country. This long extension was obviously a serious embarrassment to the Marconi Company which at once opened negotiations with Lodge for the purchase of the patent. In October 1911 an agreement was reached whereby the Lodge-Muirhead Syndicate disposed of all their patents and royalties to Marconi's for £18,000, a very considerable sum in the money-values of those days. The agreement further provided for the winding-up of the Lodge-Muirhead Syndicate and for the appointment of Lodge as Scientific Advisor to the Marconi Company for an initial term of seven years. It would be difficult to conceive of a more complete vindication of Lodge's claims in this field.

References

1. O. J. Lodge, *Past Years - An Autobiography*, Hodder and Stoughton, London, 1931, p.232..
2. O. J. Lodge, 'Protection of Buildings from Lightning', *Journal of the Society of Arts*, 1888, **36**, pp.867-874, 880-893.
3. *ibid.*, p.889.
4. *ibid.*, p.890.
5. O. J. Lodge, 'On the Theory of Lightning Conductors', *Phil. Mag.*, 1888, **26** (5th Series), pp.217-230.
6. William Thomson, 'On Transient Electric Currents', *Phil. Mag.* 1853, **5** (4th Series), pp.393-405.
7. Lodge, 'On the Theory of Lightning Conductors', p.227.
8. G. F. FitzGerald, Presidential Address to the Mathematical and Physical Section of the British Association for the Advancement of Science, 1886.
9. O. J. Lodge, 'The Discharge of a Leyden Jar', *Proceedings of the Royal Institution*, 1889, **12**, pp.413-424.
10. O. J. Lodge, 'On Electric Radiation and its Concentration by Lenses', *Phil. Mag.*, 1889, **28** (5th Series), pp.48-65.
11. O. J. Lodge, 'Easy Lecture Experiment in Electric Resonance', (letter), *Nature*, 1889-90, **41**, p.368.
12. O. J. Lodge, 'The Work of Hertz', *Proceedings of the Royal Institution*, 1894, **14**, pp.321-349. Or see *the Electrician*, 1894, **33**, pp.153-155, 186-190, 204-205. See also *Signalling Through Space Without Wires: The Work of Hertz and his Successors*, The 'Electrician' Printing and Publishing Company, London, 1898-1909 (4th edition).

13. G. M. Minchin, 'The Action of Electro-Magnetic Radiations on Films containing Metallic Powders', with a note by Professor O. J. Lodge, 'On the Sudden Acquisition of Conducting Power by a Series of Discrete Metallic Particles'. *Proceedings of the Physical Society*, 1894, **12**, pp.455-460. Also in *Phil. Mag.*, 1894, **37** (5th series), pp.90-95.

14. E. Branly, 'Variations de conductibilité des substances isolantes', *Comptes Rendus*, 1890, **111**, pp.785-787 and **1891, 112**, pp.90-93. See also: E. Branly, 'Variations of Conductivity under Electrical Influence' (a translation from *La Lumière Electrique*), *The Electrician*, 1891, **27**, pp.221-222, 448-449.

15. O. J. Lodge, 'The History of the Coherer Principle', *The Electrician*, 1897-98, **40**, pp.87-91.

16. Lodge, 'The Work of Hertz'.

17. *ibid.*

18. *The Times*, 14 and 15 August 1894. See also R. F. Pocock, 'The First Radio-Telegraph Transmission', Institution of Electrical Engineers, *Electronics & Power*, 1969, **15**, pp.327-329.

19. Lodge, *Past Years*, *op. cit.*

20. Mary Muirhead, *Alexander Muirhead*, (a biography, privately printed for the author), Basil Blackwell, Oxford, 1927, p.95.

21. H. C. Marillier, 'The Lodge-Muirhead Wireless Telegraph System', *The Electrician*, 1902-03, **50**, pp.930-934. See also J. Erskine-Murray, 'The Lodge-Muirhead System', Chapter 9 in *A Handbook of Wireless Telegraphy*, Crosby Lockwood & Son, 1907.

22. M. G. Simpson, 'Wireless Telegraphy from the Andaman Islands to the Mainland of Burma', *The Electrician*, 1906, **57**, pp.49-51.

23. *The Times*, Law Reports, 5 April 1911 (p.3), continued 6 April (p.3), and 29 April (p.28). See also S. P. Thompson, *Notes on Sir Oliver Lodge's Patent for Wireless Telegraphy* (a pamphlet, 40 pages), Waterlow & Sons, 1911.

Chapter 6

Popov

Aleksandr Stepanovich Popov (1859-1905)

We must now revert to the mid-1890s but, before considering the events which followed upon young Marconi's arrival from Italy in the early spring of 1896, it is necessary to refer to the claims which have been so widely made in recent years on behalf of Aleksandr Popov, whose work during 1895 is alleged to constitute him the 'true inventor of radio'. The claims, which have a background of political propaganda, have been disseminated so widely that it is necessary to take serious notice, even if only to demolish them.

The development of radio was, like that of so many other inventions, not the work of any single individual but the ultimate outcome of a whole succession of observations, theories and discoveries at the hands of a number of physicists and mathematicians extending over a very considerable period of time. Each stage in the progression was essential, none was redundant and each was completely dependent on what had gone before. Thus it is inappropriate, as it is with so many other inventions, to cite any particular individual as the sole 'inventor'; indeed to do so distorts the real history.

On this account it seems unfortunate that, ever since 1945, Aleksandr Popov has for internal political reasons been vigorously proclaimed by Soviet authorities as the 'inventor' of radio. Let it be said at once, therefore, clearly and categorically, that in the view of the present author, Popov made no significant contribution either in Russia or elsewhere to the development of practical radio communication. It is doubtful whether his name would even have been mentioned in these pages but for the exaggerated claims made so vociferously on his behalf by the Soviet authorities in recent years.

A. S. Popov was a lecturer in physics at the Naval Torpedo School at Kronstadt, near St Petersburg. An all-round scientist of considerable ability, he had become deeply interested in Hertzian waves after reading reports of Lodge's lectures during the summer of 1894.

It will be recalled that in June 1894 a commemorative lecture on 'The Work of Hertz' had been given by Professor Oliver Lodge at the Royal Institution. In this lecture, which had been fully and very widely reported, Lodge had not only summarized the work of Hertz himself but had amplified it with a description of his own experiments. In particular, he described 'an astonishingly sensitive detector of Hertzian waves which I call a "coherer"', and said that by means of the 'coherer', he could detect electric discharges 'at a distance of forty yards'. He added, however, 'I mention forty yards because that was one of my first outdoor experiments; I should think something more like half a mile was nearer the limit of sensitivity.'

Nowhere during the course of his lecture did Lodge express any suggestion that these new Hertzian waves might be used for signalling, nor even that they might have any other practical application. He was lecturing to physicists and his lecture was largely confined to the physical aspects of the waves, their characteristics and their identity with the waves of light.

Fascinated, as were many other physicists by the reports of Lodge's lecture and particularly by the sensitivity of his 'coherer' in detecting the occurrence of electric sparks in the vicinity, Popov began to experiment on similar lines. Setting up a circuit identical with that described by Lodge, he connected it to a lightning conductor on the roof of the Institute of Forestry in St Petersburg and found that he could use it to record the occurrence of thunderstorms at ranges of up to thirty miles. He described his results and the means he used for recording them in a long paper addressed to the *Journal of the Russian Physico-Chemical Society* in January 1896[1] which ended with the words: 'In conclusion I may express the *hope* that my apparatus, *when further perfected* may be used for the transmission of signals over a distance with the help of rapid electric oscillations, *as soon as a source of such oscillations posssessing sufficient energy will be discovered*' [present author's italics].

In the words of Professor Ambrose Fleming a few years later:

> We are left, then, with this unquestionable fact that at the beginning of 1896, although the most eminent physicists had been occupied for nine years in labouring in the field of discovery laid open by Hertz, and although the notion of using these waves for telegraphy had been clearly suggested, no one had overcome the practical difficulties, or actually given an exhibition in public of the transmission of intelligence by alphabetic or telegraphic signals by this means. The appliances in a certain elementary form existed, the advantages and possibilities of electric wave telegraphy had been pointed out, but no one had yet conquered the real practical difficulties, and exhibited the processes in actual operation.[2]

It is quite evident, then, that in January 1896 (a few months before Marconi demonstrated his system) Popov did *not* possess a practicable method of radio-communication. He was merely expressing the 'hope' that, when further developed, his apparatus might be used for the reception of signals, but even this, he was admitting, would have to await the arrival of a sufficiently powerful transmitter (actually, this judgement was wrong: what was really needed at this juncture was not a more powerful transmitter, but a more sensitive receiver). As

Fleming has so accurately recorded, 'at the beginning of 1896, the means existed, the possibility of communicating without wires had been expressed by a number of writers but no one had as yet put the pieces together and demonstrated the process in actual operation'. Popov had *not* invented radio.

Why, then, did the Soviet authorities so vociferously proclaim that he had? The answer almost certainly lies in the chaotic aftermath of the Second World War (1939-1945) when, in the east, Stalin's forces had finally triumphed over the Nazis but only at the cost of enormous losses in lives and materials. Victory had only been won with the aid of the vast material and technological assistance provided by the USA and the United Kingdom. In earlier years, scientific research and technological development had received insufficient attention in Russia from official sources and in these respects the Soviets had lagged behind western nations. In the post-war era, however, the authorities considered it necessary on political grounds to conceal their deficiencies from the common people and a nation-wide propaganda campaign was instituted to enhance the reputations and achievements of former Russian scientists. Nationalistic enthusiasms prevailed and exaggerated claims were made.

Popov was not alone: over several years a number of former Russian scientists, physicists and engineers were similarly paraded as the original pioneers or inventors in their particular fields - even television and the aeroplane were similarly claimed to be Russian inventions - all with the object of concealing from the common citizen the fact that Russian science had been allowed to lag behind developments in the western world.

Deeply concerned at the discrepancy between the claims made on behalf of Popov and the more usually accepted versions of early radio history, some years ago Professor Charles Susskind of the University of California carried out an exhaustive enquiry into the Soviet claim that Popov was the original 'inventor' of radio. The results of Susskind's very thorough research into all the contemporary records were presented in a long paper[3] addressed to the Institute of Radio Engineers in 1962. He says, 'The argument turns principally on whether Popov used his instrument merely to register lightning flashes and their man-made equivalents, or whether he did actually publish, before mid-1896, a description of the use of his instrument for the transmission of intelligence. *The record shows that he did not.*' Susskind's paper concludes:

> The Russians have good reason to be proud to have produced a pioneer of Popov's rank; but the officious Soviet campaign to designate him the 'inventor of radio' and to enlarge his reputation out of proportion with his achievements amounts to a deviation from objectivity that must be deplored by all historians of technology who remain untouched by chauvinistic considerations.
>
> The entire world has been impressed by the recent technical achievements of the Soviet peoples; as they near the front rank in the march of modern technology, they can well afford to abandon the unbecoming and recurrent protestations of priority characteristic of an earlier era when they were further

behind and their leaders considered it necessary to carry over the personality cult from the political sphere into the field of modern invention.

References

1.   A. S. Popov, *Zh. Russ. Fiz.-Khim. Obshchestva* (Physics, pt 1) 1896, **28**, pp.1-14.
2.   J. A. Fleming, *The Principles of Electric Wave Telegraphy*, Longmans, Green & Co., 1906. pp.361, 425-426.
3.   Charles Susskind, 'Popov and the beginnings of Radiotelegraphy', *Proceedings of the Institute of Radio Engineers*, 1962, **50**, pp.2036-2047.

Chapter 7

Marconi

Guglielmo Marconi (1874-1937)

The story of Marconi's birth on 25 April 1874 and his upbringing in Italy has been told by others, notably by Professor W. P. Jolly in his biography, *Marconi*,[1] and there is no need to repeat it here. It is sufficient to note that he had commenced his experiments during the autumn of 1894 and that during the summer of 1895 he had been able to send signals on his father's estate over a distance of about one mile. While the facts relating to his boyhood are well-known, the story of his first few years in England is virtually a 'closed book'. Many writers have attempted to tell the story, but the almost complete absence of any documentary evidence or contemporary records has resulted in the early history of radio communication largely consisting of a few salient facts imaginatively cobbled together but totally lacking substantive detail to authenticate and verify the chain of events.

Even the 'accepted' story that Marconi was given an introduction by A. A. Campbell-Swinton, one of the leading consulting electrical engineers of the period, to Sir William Preece, Chief Engineer to the Post Office, and that everything followed from this introduction can now be shown to be only half the truth. Marconi's original approach was made not to the Post Office but to the Secretary of State for War. The invention he was offering in his original letter, which is reproduced overleaf, has turned out not to be for a system of communication but for a system for the radio-control of torpedoes or other unmanned vessels! However, it was the keen perception of the officer on the staff of the War Office deputed to enquire into the matter, Major C. Penrose, which proved a major influence in persuading government departments to take the young man seriously.

Patient research during recent years, coupled with the fortunate discovery of many of the original files and reports in the archives of the Post Office, the Admiralty and the Public Record Office, has now made it possible to tell the full story of Marconi's early years in England and, for the first time, to cloak it with a wealth of fascinating detail.

My Lord

I Guglielmo Marconi of Bologna (Italy) now residing at 71 Hereford Road Bayswater W London do hereby declare that I have discovered electrical devices which enable me to guide or steer a self propelled boat or torpedo from the shore or from a vessel without any person being on board the said boat or torpedo.

It is not necessary to have any communication whatever such as wires or ropes between the self propelled boat or torpedo and the person directing its evolutions.

I have found it possible by turning the handle of a simple apparatus of my invention to turn about steer or enable the independent boat or torpedo to pursue any object at more than a mile from the shore or ship from which it has been launched provided the boat or torpedo has an apparatus of my invention applied to its rudder.

Should Your Lordship consider my invention useful to Her Majesty's Army or Navy I am willing to demonstrate its practicability at my own expense, by means of a small self propelled boat on any lake or river where Your Lordship may desire.

I humbly beg Your Lordship to honour me as soon as possible with an answer as I propose otherwise to obtain patents for my discovery, and apply it to commercial purposes.

I have the honour to be my Lord

Your most obedient servant
Guglielmo Marconi

71 Hereford Road
Bayswater W London

20th May 1896

To the Right Honourable Lord
Her Majesty's Principal Secretary of State
for War Affairs

FOTO UNGARIA , ROMA

A. DELZERS

GUGLIELMO MARCONI

1874 - 1937

Figure 10   Guglielmo Marconi

Accompanied by his Irish mother, Marconi arrived in London towards the end of February 1896, just two months before he was due to celebrate his twenty-second birthday. They took up residence initially in a private hotel at 71 Hereford Road, Bayswater, but moved a few months later to 21 Burlington Road, St Stephen's Square, and again, early in 1897, to 67 Talbot Road, Westbourne Park.

Following his arrival in London, Marconi lost little time in making his first application for a patent to protect his invention. It was filed at the Patent Office on 5 March 1896 and entered in the Official Journal as No. 5028, with the title *Improvements in Telegraphy and in apparatus therefor*. It seems not improbable that Marconi had drafted his application while still in Italy and that he subsequently found it to be inadequate but, whatever the cause, he abandoned this first application. Since it is the practice of the Patent Office after a period to destroy abandoned applications, no details of the application other than the title have survived. Its abandonment may account for historians having overlooked the fact that it was ever made, Marconi normally being credited with having made his first application on 2 June 1896. This latter application was drafted by Fletcher Moulton, QC, a distinguished patent agent of the time, and accepted as No.12,039/1896; it was the first patent relating to wireless telegraphy ever granted.

Marconi had been welcomed on his arrival in London by his cousin, Colonel Henry Jameson-Davis (a son of Mrs Marconi's elder sister), who was established in London, at 12 Mark Lane, as a professional engineer and an authority on the design and construction of flour-milling machinery. Colonel Jameson-Davis' guidance and advice to his young cousin was to prove invaluable - indeed it may well have been crucial. His connections and experience in the professional and business worlds more than compensated for the obvious deficiencies in his young cousin from Italy. Marconi himself was not lacking in self-confidence, but in 1896 it required more than personality to gain access to and a hearing by professional engineers and senior government officials. It was Colonel Jameson-Davis who provided the key to open the doors but, once inside, Marconi's own personality and obvious ability soon convinced his hearers that he had something to offer and he persuaded the authorities to arrange trials first on Salisbury Plain and then at Lavernock in the British Channel.

According to J. Gavey, the Engineer-in-Chief of the General Post Office in November 1902, the results of these tests in the Bristol Channel gave conclusive proof of the inherent possibilities of the system.

Figure 11   Sir William H. Preece

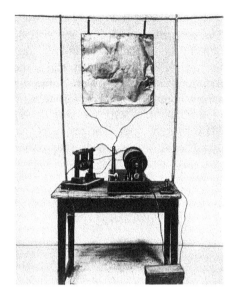

Figure 12   A replica of Marconi's first wireless transmitter as used by him in 1895-6

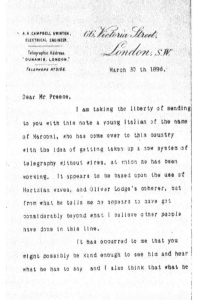

Figure 13   Marconi's letter of introduction from A. A. Campbell-Swinton

*The Lavernock trials**

The experiments that were carried out by Marconi on Salisbury Plain in September 1896 and March 1897 demonstrated that his system of signalling by the use of Hertzian waves had some potential value but the results, although extremely interesting, were somewhat tentative. They had, however, aroused serious interest in the Post Office, and it was therefore decided to institute a series of official trials with the specific object of ascertaining the practical value of Marconi's methods. The results of the trials were of immense importance, yet the details have not been published previously.

Figure 14 Impression of Marconi's demonstration on Salisbury Plain, 1896, by the artist Steven Spurrier

The trials were held in May 1897 between Lavernock Point, about five miles south of Cardiff, and the island of Flat Holme in the Bristol Channel, a distance of about 3.3 miles. This location was chosen partly because it provided a convenient site for testing transmission over water but mainly because it had been the site where Preece's 'parallel-wire system' had been tested several years earlier. This was a system for signalling across stretches of water, by means of

---

* The account in the following pages was written by the present author and first published in *Electronics & Power*, 2 May 1974.

Figure 15 Post Office staff examining Marconi's apparatus for experiments across the Bristol Channel, 1897. *L to r:* S. E. Hailes (linesman), H. C. Price G. N. Partridge (engineers)

conduction, in circumstances where the laying of a cable was not practicable, as between the shore and a lighthouse. It now provided a reliable means of communication between Lavernock Point and Flat Holme island while tests with the somewhat fickle Marconi apparatus were in progress.

The Lavernock trials were in every sense official Post Office trials, carried out principally by Post Office staff, with the close co-operation and goodwill of Signor Marconi who, at the time, was only just twenty-three years of age.

Prior to the commencement of the tests, a pair of masts about 130 feet (35 m) high had been erected at Lavernock Point and on Flat Holme. At that time it was believed that a substantial area of capacity was a desirable, if not an essential feature, of a wireless aerial and large drums of sheet zinc, suitably supported and insulated, were therefore installed at the top of each mast. According to Fahie,[2] these 'elevated capacities' were cylindrical caps, 6 feet (1.8 m) long and 3 feet (0.9 m) in diameter and they were connected to the signalling apparatus at the foot of each mast by substantial strands of gutta-percha covered aluminium wire. A number of thick stranded copper wires leading over the edges of the cliffs into the sea made the 'earth' connections.

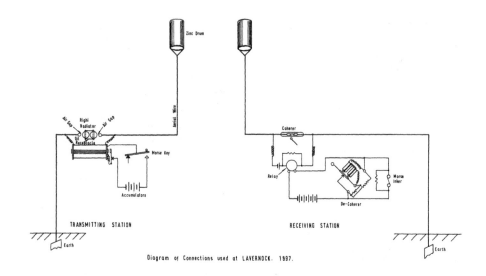

Diagram 6 Transmitter and receiver circuit used during Lavernock trials.
From Engineer-in-Chief's report, GPO 1903

The apparatus used throughout the Lavernock trials was partly Marconi's own equipment and partly modified and redesigned gear made up in the Post Office workshops but, inevitably at that early stage, much was of a crude nature that would need developing to give a reliable system for practical use.

The transmitter on Flat Holme employed a Rühmkorff induction coil connected to a Righi oscillator, a device peculiarly unsuitable for the frequencies Marconi was now using. Briefly, the Righi oscillator comprises a set of four brass spheres, the two central ones being about 4 inches (10cm) in diameter and separated by a gap of approximately 1 to 1.5 millimetres, while the two outer spheres were each about 1 inch (2.54 cm) in diameter and separated from the respective inner balls by a gap of about 0.5 inch (1.27 cm). The secondary winding of the induction coil was connected to the two outer spheres, so that the spark discharge had to bridge three gaps in the aerial-earth system. Three different coils were used during the experiments. The first, a powerful instrument capable of giving a 20-inch (51 cm) spark, had been specially purchased by the Post Office. The second was a small instrument of similar pattern, giving a 10 inch (25 cm) spark. The third, the property of Signor Marconi, was of intermediate size, but it became defective early in the tests so the main work was carried out with the Post Office coils.

From the several available sources it is possible to piece together a fairly complete list of those who attended the trials during at least part of their duration.

Among the Post Office officials were Mr (later Sir) William Preece, then Engineer-in-Chief, John Gavey, a Principal Technical Officer and Preece's successor, Mathew Cooper and J. E. Taylor.[3] The War Office was represented by Major C. Penrose, and the Royal Engineers by Major G. A. Carr and Captain J. N. C. Kennedy. Professor Viriamu Jones of University College, Cardiff, was an 'interested spectator'. The one whose presence was to cause most trouble for the future Marconi Company was Professor A. Slaby of Charlottenberg, who had been reluctantly invited by the Post Office at the personal request of the German Emperor.

A name that does not appear in any of the contemporary accounts is that of G. S. Kemp, to whom we are chiefly indebted for the day-to-day account of the Lavernock trials. Kemp had joined the Royal Navy at the age of fifteen and, when he was discharged in 1895 at the age of thirty-eight, he joined the staff of the Post Office as an assistant in the Engineer-in-Chief's Laboratory. In that capacity, he had been instructed to assist Marconi in the earlier experiments on Salisbury Plain. With the decision to hold the more extensive trials in the Bristol Channel, Kemp was made responsible for transporting and setting up the apparatus at Lavernock and Flat Holme. Following his life-long habit, Kemp recorded brief details of his daily activities in a pocket diary. It is to this diary - and particularly to the expanded and edited versions that Kemp prepared for the Marconi Company in about 1930 - that we are indebted for the details which follow.

In Kemp's words, the historic experiments started early:

Thursday, 6 May
Left at 8.30 a.m. for Paddington with apparatus for experiments at Cardiff. Arrived at 2.17 p.m. and stowed apparatus in store.
Proceeded to Lavernock to see mast and found that a long cable had been fixed, stretching out beyond low-water mark, for the earth connection. Fixed a wire atop the 107 ft pole, 16 strands of aluminium wire. Then returned to Cardiff to make arrangements for transporting apparatus to Flat Holme Island.

Friday, 7 May
I packed Mr Marconi's transmitter into a small tug at 6.30 a.m. together with the transmitting and receiving apparatus belonging to Mr Preece's Parallel-Wire system and transported all to Flat Holme Island. Fixed a wire of 18 strands to top of 110 ft pole and prepared Mr Marconi's transmitter in a small hut close to mast. Slept at a small house owned by the person in charge of the Cremation House.

For the next few days, Kemp was busy on the little island, fitting up and testing Marconi's transmitter and Preece's parallel-wire system. He had trouble with the insulation of the zinc drum at the top of the mast and with the insulation of the stays. Sparks on the parallel-wire system also caused difficulties whenever he used the Marconi transmitter, but these were only 'teething troubles' and by the Wednesday of the following week he was able to record that, 'The signals transmitted across to Lavernock by Mr Marconi's transmitter and the Parallel-

Wire system were good.' Insulation, however, was still proving troublesome and his next comment was, 'As I did not like the insulation of the drum, I sent some of these signals on the aerial which was connected to insulated stays' - a reminder of the very high voltages encountered in the aerial circuits of the early spark transmitters.

Mention was made above of the two versions of Kemp's diary: the original contemporary pocket diary[4] (parts of which the owner, Kemp's son, Leslie, kindly permitted me to photograph some years ago) and the expanded version which Kemp had typed and edited for the Marconi Company in about 1930. The latter version contains an amount of detail to which no reference is made in the original, and while no actual contradictions have been noted, it is difficult to avoid wondering how an old man (he was over seventy at the time) writing more than thirty years after the events could have remembered many of the trivial details he mentions. In the original diary, the events of the time from Monday 10 to Friday 14 May are bracketed together with the single comment, 'Signalling with Marconi and Parallel-Wire systems to Lavernock Point', but in the 1930 version the daily events are recorded with considerable detail, for example:

Thursday, 13th May
The great day for Flat Holme signals. I started at 7 a.m. and fitted a new copper earth wire in lieu of the iron earth. I sent and received good signals on both systems between 12 and 1.45 p.m. The first half hour of V's were on a paper strip on the inker, the second, 'so be it, let it be so', and the third, 'it is cold here and the wind is up'. This message was posted to the Kaiser by Professor Slaby.
In the afternoon Mr Marconi came over and tried some adjustments; Mr Taylor came with him and did a little transmitting but, as I sent the best sentences between 12 and 2 p.m. I returned to those adjustments and sent them the following:

How are you?	repeated
It is hot	repeated
Marconi	repeated
Go to bed	repeated
Go to Hull	repeated
So be it	repeated
Tea here is good	repeated

Nine similar sentences follow. The tests were resumed the following morning. A motor-driven commutator and a Vrill break were tried, but with no marked improvement on the previous day's results.

Saturday, 15th May
I dismantled the Marconi transmitting apparatus on Flat Holme, leaving it at Penarth, and then arranged for a steamer to Brean Down on Monday.

Here is the first mention in any of the records of an attempt to transmit right across the Bristol Channel from Lavernock to Brean Down on the Somerset coast.

Figure 16   George Kemp with one of the signal kites, photographed at the Science
Museum

It leaves the impression that it was a sudden, 'on the spot' decision, inspired in all probability by the success of the Lavernock-Flat Holme experiments. Preparations continued over the weekend, with Kemp assembling the Marconi transmitter on the top of the cliff at Lavernock. Monday, however, brought bad weather, and Kemp noted that it was too rough for the receiver party to land at Brean Down. Kemp himself remained at Lavernock to operate the transmitter. Just how the receiver party eventually reached Brean Down is not evident from the surviving records. There is no record either of the names of those in the receiver party, or of exactly what they received, but in his contemporary diary Kemp noted on Tuesday, 18 May: 'Good signals to Brean Down using kite and 300 ft (91.4m) of 4-strand wire'.

In the language of the day, the phrase 'good signals' was far from being synonymous with 'good messages'. In the 1930 version of his diary, Kemp seems to qualify his original comment by saying 'The engineers reported that they had received signals at Brean Down'. Whether or not the signals were exactly 'Q5' (fully readable), it is evident from Gavey's report[5] that the Post Office officials were impressed with the inherent possibilities of the system. Signals, of sorts, had got across, although it was clear that, in Gavey's words, 'There was . . . still much to be desired in order to convert crude appliances into good working devices'.

In fact, the apparatus that Marconi was using at the time of the Lavernock trials possessed two fundamental faults, one in the transmitter and the other in the receiver, and it is perhaps surprising that the results obtained were as good as they were. At the transmitter, the frequency of the principal oscillations would have been determined by the diameter of the two large brass balls of the Righi oscillator at approximately 500-800 MHz. But the two outer balls were connected to the aerial and earth respectively, and this system, while the spark gaps were sufficiently ionized, would have possessed a resonant frequency of approximately 2-3 MHz. Inevitably, the system would have been very inefficient.

Similarly at the receiver a fundamental error was involved by interposing the coherer directly in the lead connecting the aerial and earth. In its sensitive condition, a coherer possesses a very high resistance, and by locating it directly in the aerial-earth circuit, it clearly interrupted the continuity of the oscillatory system. While a number of factors militated against this error, it undoubtedly contributed significantly to the capricious and inconsistent results that were so often experienced.

This historic series of experiments across the Bristol Channel came to a close, as Kemp noted the following in his diary:

Saturday, 29th May
Packed up and returned to Paddington by the 10.37 p.m. train from Cardiff, arriving at Paddington at 3.30 a.m. on Sunday morning. We stowed all the apparatus in the cloak room.

However, this was not quite the end of the association of Lavernock and Brean Down with the early development of wireless telegraphy.

Only a few weeks after the completion of the trials, Marconi and his associates registered the Wireless Telegraph and Signal Company, a step that unfortunately led to the Post Office breaking off all collaboration with Marconi. The Secretary of the Post Office and his legal advisers took the view that it would be improper for the Post Office to spend further time and money in developing an invention, the patent rights in which were held by a public company. As a result, the Engineer-in-Chief was formally instructed that 'for the present Mr Marconi could not take part in any Post Office experiments whatever'.

For almost two years the subject of wireless telegraphy was dropped completely within the Post Office, but by the autumn of 1899 it became known that Marconi was achieving ranges of approximately sixty miles, and it seemed desirable for the Post Office to pick up the threads once more. An immediate incentive was provided by an approach from an Hungarian inventor, Bela Schaefer, who offered a receiver that might provide an alternative to that described in the Marconi patent. Arrangements were made accordingly to test the Schaefer detector and compare it with the coherer method of reception in a further series of experiments across the Bristol Channel between Lavernock and Brean Down.

In the event, the instability of the Schaefer detector showed it to be quite useless for the purposes of communication, but the Post Office now began to pursue the subject with determination. The Brean Down site was soon found inconvenient for protracted experiments, and the station was removed to a more sensible location in the bay of Weston-super-Mare. With only brief intervals, experimental transmissions were carried on between Lavernock and Weston from October 1899 to April 1900. By the time they were concluded, the apparatus and circuitry had been much improved, so that regular and fairly easy communication had become possible.

Deeply regrettable though the rift with Marconi in the autumn of 1897 may have been (especially in the way it clouded relations between the Post Office and the Marconi Company for many years to come), there can be little doubt that the two series of Lavernock trials and the associated events were of the utmost importance in leading to the establishment within the Post Office of a strong department dealing particularly with wireless telegraphy. Indeed, it is to the events related above that the whole of modern radio can trace its foundation.

References

1. W. P. Jolly, *Marconi: a biography,* Constable & Co, London, 1972.
2. J. J. Fahie, *History of Wireless Telegraphy,* Blackwood, 1900. See also J. J. Fahie, *A History of Wireless*, Blackwood & Sons, 1900.
3. J. E. Taylor, 'The Early History of Wireless Telegraphy', *Post Office Electrical Engineers Journal,* 1909, 2, pp.157-159.
4. G. S. Kemp, Contemporary manuscript diary originally in the possession of Leslie Kemp (now deceased).
5. J. Gavey, Report by Engineer-in-Chief of the General Post Office, on 'Technical Aspects of Wireless Telegraphy', HMSO, 1903.

Index

Distant magnet 9
Doll, Dr. Max (Geometry Lecturer at
    Technical High School, Karlsruhe)
    44
Doll, Elizabeth (Dr. Doll's daughter and
    Hertz' future wife) 44
Dynamical theory 25, 27

Ebert, Professor Hermann 46
Ecole Polytechnique 3
Electrician, The 65, 68
Electrostatic field 21
Electrostatic forces 21
Electrostatic induction 15
Electrotonic state 14–15
Ether 16–19, 22, 26, 31
Extra current 28, 42

Fahie, J. J. 29, 81
Faraday, Michael 1, 4–19, 21–22, 34
Fedderson 45
Field theory 2, 6, 11, 34
FitzGerald, George Francis 30–32, 43,
    54, 58
Fizeau, A. H. L. (Physicist) 25
Flat Holme Island 80–84, 86
Fleming, Professor Ambrose (Physicist)
    72–73
Fletcher Moulton, Q.C. (Patent agent)
    78
Foster, Professor Carey (University
    College, London) 53–54
Frankfurt-am-Main 38–39
Frankfurt Physics Club 39
Frankland, Professor (Sir Edward F.,
    Professor of Chemistry) 53
Franklin, Benjamin (Scientist) 55
French Academy 3
Fresnel, Augustin (Physicist) 9

Galvanic battery 12
Galvanic current 13
Galvanic electricity 14
Galvanometer 3, 9, 66
Gavey, John (Post Office Engineer-in-
    Chief) 78, 83, 86
Geissler tubes (Heinrich Geissler) 26
Gewerbeschule (Technical School) 37

Glenlair 27
Greely, General (USA Signal Corps) 67
Griefswald 44
Gutta percha 81

Hamburg 36, 38
Hamburg Gymnasium 38
Hamilton's quaterionic calculus
    (Sir William R. Hamilton) 26
Heaviside, Oliver (Physicist) 57
Helmholtz, Professor Herman von
    (Physicist) 27, 30, 41–45, 48, 54
Henrichi, Professor (University College,
    London) 53
Henry, Joseph 11–13, 54–55, 60
Hertz dipole 46–48, 61–62
Hertz, Elizabeth 49
Hertz, Dr. G. F. 36–37
Hertz, Gustav 36
Hertz, Heinrich 1, 5, 27–28, 32, 34–49,
    54, 58, 60–61, 65, 67, 72
Hertz, Johanne 49
Hertz, Mathilde 49
Hertz, Melanie 36
Hertz, Otto 36
Hertz, Rudi 36
Hertz oscillator (Primary conductor)
    47, 61, 63, 65
Hertzian waves 64–67, 71–72, 80
High frequency rectifier 28
High voltage electricity 14
History of radio 37
Howard, Dr. (Oliver Lodge's assistant)
    61
Hughes, David Edward 28–30, 35
Huxley, Professor 28–29, 53, 59

Induction balance 28
Induction coil 43, 45, 47, 62, 66, 82
Induction of static electrical charges 16
Inductive circuit 28
Institute of Forestry, St. Petersburg 72
Institute of Radio Engineers 73
Intermolecular currents 4

Jameson-Davis, Colonel Henry
    (Marconi's cousin) 78

Printed in the USA
CPSIA information can be obtained
at www.ICGtesting.com
JSHW011816301024
72690JS00002B/102